Building the Hyperconnected Society

IoT Research and Innovation Value Chains, Ecosystems and Markets

RIVER PUBLISHERS SERIES IN COMMUNICATIONS

Volume 43

Series Editors

ABBAS JAMALIPOUR
The University of Sydney
Australia

MARINA RUGGIERI
University of Rome Tor Vergata
Italy

HOMAYOUN NIKOOKAR
Delft University of Technology
The Netherlands

The "River Publishers Series in Communications" is a series of comprehensive academic and professional books which focus on communication and network systems. The series focuses on topics ranging from the theory and use of systems involving all terminals, computers, and information processors; wired and wireless networks; and network layouts, protocols, architectures, and implementations. Furthermore, developments toward new market demands in systems, products, and technologies such as personal communications services, multimedia systems, enterprise networks, and optical communications systems are also covered.

Books published in the series include research monographs, edited volumes, handbooks and textbooks. The books provide professionals, researchers, educators, and advanced students in the field with an invaluable insight into the latest research and developments.

Topics covered in the series include, but are by no means restricted to the following:

- Wireless Communications
- Networks
- Security
- Antennas & Propagation
- Microwaves
- Software Defined Radio

For a list of other books in this series, visit www.riverpublishers.com
http://riverpublishers.com/series.php?msg=Communications

Building the Hyperconnected Society

IoT Research and Innovation Value Chains, Ecosystems and Markets

Editors

Dr. Ovidiu Vermesan

SINTEF, Norway

Dr. Peter Friess

EU, Belgium

River Publishers

Published, sold and distributed by:
River Publishers
Niels Jernes Vej 10
9220 Aalborg Ø
Denmark

ISBN: 978-87-93237-99-5 (Hardback)
     978-87-93237-98-8 (Ebook)
©2015 River Publishers

Dedication

"The greatest accomplishments of man have resulted from the transmission of ideas and enthusiasm."

— Thomas J. Watson

"A rock pile ceases to be a rock pile the moment a single man contemplates it, bearing within him the image of a cathedral."

— Antoine de Saint-Exupéry

Acknowledgement

The editors would like to thank the European Commission for their support in the planning and preparation of this book. The recommendations and opinions expressed in the book are those of the editors and contributors, and do not necessarily represent those of the European Commission.

Ovidiu Vermesan
Peter Friess

Contents

Preface

Internet of Things beyond the Hype

IoT represents the convergence of advances in miniaturization, wireless connectivity, increased data storage capacity and data analytics. Intelligent edge devices detect and measure changes in environmental parameters and are necessary to turn billions of objects into "smart data" generating "things" that can report on their status, and interact with other "things" and their environment.

Universal connectivity and data access provides opportunities to monetise data sharing schemes for mobile network operators and other connectivity players.

The Internet of Things supports private and public-sector organizations to manage assets, optimize performance, and develop new business models, allowing a leap in productivity while reshaping the value chain, by changing product design, marketing, manufacturing, and after sale service and by creating the need for new activities such as product data analytics and security. This will drive yet another wave of value chain based productivity improvement.

The following chapters will provide insights on the state-of-the-art of research and innovation in IoT and will expose you to the progress towards building ecosystems and deploying Internet of Things technology for various applications.

Editors Biography

Dr. Ovidiu Vermesan holds a Ph.D. degree in microelectronics and a Master of International Business (MIB) degree. He is Chief Scientist at SINTEF Information and Communication Technology, Oslo, Norway. His research interests are in the area of microelectronics/nanoelectronics, analog and mixed-signal design with applications in measurement, instrumentation, high-temperature applications, medical electronics, integrated intelligent sensors and computer-based electronic analysis/simulation. Dr. Vermesan received SINTEFs 2003 award for research excellence for his work on the implementation of a biometric sensor system. He is currently working with projects addressing nanoelectronics integrated systems, communication and embedded systems, wireless identifiable systems and cyber-physical systems for future Internet of Things architectures with applications in green automotive, internet of energy, healthcare, oil and gas and energy efficiency in buildings. He has authored or co-authored over 75 technical articles and conference papers. He is actively involved in the activities of the new Electronic Components and Systems for European Leadership (ECSEL) Joint Technology Initiative (JTI). He coordinated and managed various national and international/EU projects related to integrated electronics. Dr. Vermesan is the coordinator of the IoT European Research Cluster (IERC) of the European Commission, actively participated in projects related to Internet of Things.

Dr. Peter Friess is a senior official of DG CONNECT of the European Commission, taking care for more than six years of the research and innovation policy for the Internet of Things. In his function he has shaped the on-going European research and innovation program on the Internet of Things and accompanied the European Commission's direct investment of over 100 Mill. Euro in this field. He also oversees the international cooperation on the Internet of Things, in particular with Asian countries. In previous engagements he was working as senior consultant for IBM, dealing with major automotive and utility companies in Germany and Europe. Prior to this engagement he worked as IT manager at Philips Semiconductors on with business process optimisation in complex manufacturing. Before this period he was active as researcher in European and national research projects on advanced telecommunications and business process reorganisation. He is a graduated engineer in Aeronautics and Space technology from the University of Munich and holds a Ph.D. in Systems Engineering including self-organising systems from the University of Bremen. He also published a number of articles and co-edits a yearly book of the European Internet of Things Research Cluster.

1

Introduction

Thibaut Kleiner[1]

European Commission

1.1 Now Is the Time

In 1999, Kevin Ashton coined the term Internet of Things (IoT) to describe an evolution of the Internet whereby we *'empower computers with their own means of gathering information, so they can see, hear and smell the world for themselves, in all its random glory.'*[2] At that time, it was already clear that IoT is more than a technology (and definitely more than RFID), and that it represents a paradigm, a new stage of evolution for the Internet. The EU embraced it in 2009 with a dedicated action plan, leading notably to the creation of the European Internet of Things Research Cluster (IERC).

Over the years, however, the very concept of IoT has seemed to lose traction and to become blurred, especially as a series of corporate actors have tried to develop new terminologies – from Internet of Everything to Industrial Internet to Industrie 4.0- to explain how they would deliver better solutions on the basis of connected devices. Time has come to reclaim some ground, and to re-establish the Internet of Things where it belongs: as the leading paradigm to describe the digital transformation of our economies and societies.

The IoT is *the* key development to Building the Hyperconnected Society-the topic of this book. The European Commission has adopted on 6 May 2015 the Digital Single Market strategy and has opened the door for bold proposals to improve our future. Today, we can mobilise the important research work delivered notably by the IERC in terms of IoT technology and societal analysis, and apply it in the market and in our EU policies. The launch of

---

[1]The views expressed in this article are purely those of the author and may not, in any circumstances, be interpreted as stating an official position of the European Commission.

[2]K. Ashton. That 'Internet of Things' Thing. *RFID Journal.* www.rfidjournal.com. June 22, 2009.

the Alliance for IoT Innovation (AIOTI) should be seen as a signal in this direction. We are uniquely positioned to choose the right path so that the IoT can be mainstreamed, so that it leaves the labs and the drawing boards, conquers not only the techno-freaks and early adopters but can also be adopted by the masses in full confidence.

1.2 The Digital Single Market and Internet of Things Transformative Technologies

The Internet of Things has long been characterised as hype. Already in 2009, the Commission explained that the scope of IoT applications is expected to greatly contribute to addressing today's societal challenges, from health monitoring systems to transport to environment, gradually resulting in a genuine paradigm shift. Progress has been constant but maybe slower than anticipated. This is however changing, as highlighted in a recent study completed for the Commission[3], which forecasts the market value of the IoT in the EU to exceed one trillion euros in 2020.

The Digital Single Market (DSM), adopted in May 2015, offers an opportunity to accelerate and to fully develop the transformative potential of the IoT. It announces a series of initiatives that together can boost take up on a continental basis. First, a revamped telecom regulatory framework will provide improved rules on e.g. roaming, net neutrality and spectrum and help the deployment of connected devices and IoT services. The DSM also consolidates initiatives on trust and security and data protection, which are essential for the adoption of this technology. Most importantly, it announces an initiative on the Data economy (free flow of data, allocation of liability, ownership, interoperability, usability and access) and promises to tackle interoperability and standardisation.

Altogether, these measures offer a fantastic platform to establish the framework conditions for a vibrant development of the IoT in the EU. This comes at a fruitful moment, when powerful demand forces led by socio-demographic trends, government initiatives and the expanding consumer market are driving growth in the market.

As flagged by the DSM, the main emerging markets in the short-medium term will be characterised by a combination of IoT with Cloud Computing

---

[3]IDC, TXT. *Definition of a Research and Innovation Policy Leveraging Cloud Computing and IoT Combination.* European Commission, SMART number 2013/0037. 2015.

and Big Data. In sight is the emergence of "smart environments" where hyper-connectivity and data intelligence generate multiple new services (also with other technologies such as robotics) and improve not only efficiency but also spurs innovation, increasing quality of life and tackling societal challenges. Areas like smart cities, smart homes, smart grid and smart mobility are already witnessing the emergence of new ecosystems for IoT solutions, applications and services. The transformation power goes beyond new actors and is likely to touch the core activities of established players too, as highlighted by the European Round Table of Industrialists[4]. This is an opportunity but also a risk for incumbents if they do not adjust fast enough.

1.3 Benefits and Challenges

Now that digitisation is progressing and IoT is affecting increasing numbers of companies in different areas, new questions are emerging for the ability of the EU to benefit from this process.

First, there are still some fundamental design questions, in terms of how the IoT technologies will be organised and structured. Admittedly, we may have all building blocks: smaller, lighter, more power-efficient, and cheaper hardware, more intelligent sensors and actuators, new platforms, ubiquitous wireless connectivity, available cloud services and data analytics tools. But the IoT is still characterised by vertical silos, which limits the creation of vibrant ecosystems. PWC identifies a series of obstacles in that context: market fragmentation, lack of unified standards and coexistence of open and proprietary solutions, vertical focus[5].

Another major challenge is the lack of an established horizontal platform[6] that is pervasive enough to structure and nurture the IoT ecosystem. Whilst some IoT solutions will certainly remain vertically orientated (e.g. to address mobility or healthcare needs), the highest degree of innovation is expected to be across areas (ex: car and home and city). But it is unlikely that IoT solutions can be economically developed across different areas without horizontal

---

[4]Press statement on Digitisation by Benoît Potier, chairman of the European Round Table of Industrialists (ERT) at the meeting with Chancellor Merkel, President Hollande and President Juncker, 1 June 2015.

[5]PWC. *IoT Benmark Study*. European Commission. 2015

[6]A platform can be defined as comprising the hardware (including computing and storage), software, communications, management (of the above and of intelligent and/or embedded systems), orchestration, and services (data, APIs, analytics, etc.).

platforms enabling core service elements to be managed across verticals and companies.

We will need to avoid the same result as what happened for the mobile ecosystems, where the leading platform providers are not headquartered in Europe. The DSM is launching an investigation process into platforms, which could also be relevant for the IoT, even though one should not confuse the regulatory oversight of platforms with the development of new ones. The challenge for the EU is to develop these platforms independently.

In that context, on-going efforts through Horizon2020 and dedicated research calls around open platforms for develop IoT ecosystems, and large scale IoT pilots for real-life experimentation, have the potential to help establishing the EU at the forefront of a massive deployment of the IoT, and one that is endorsed by EU citizens.

1.4 Conclusion

For many years, debate around the IoT has evolved between technology explorations and philosophical and ethical conjectures, to the point that it could jeopardise the business appetite for engaging in this research agenda. Fortunately, this exploratory stage is being superseded by a new appetite for growing the IoT market. Past debates and research findings have not been lost. They should now be mobilised to speed up the market uptake and to address the important remaining issues that may hamper the mainstreaming of the IoT. The European Commission will support this agenda.

2

New Horizons for the Internet of Things in Europe

Peter Friess and Rolf Riemenschneider

European Commission, Belgium

2.1 Introduction

The Internet of Things (IoT) represents the next major economic and societal disruption enabled by the Internet, and any physical and virtual object can become connected to other objects and to the Internet, creating a fabric between things as well as between humans and things. The IoT offers to merge the physical and the virtual worlds into a new smart environment, which senses, analyses and adapts, and which makes our lives easier, safer, more efficient and user-friendly.

Originally, the Internet was conceived to interconnect computers and transmit messages with limited data exchange capability. With the advent of web technologies, a first revolution took place enabling the linking of documents and the creation of a world wide web of information (Web 1.0). In the early 2000, the Internet evolved towards a universal communication platform making it possible to carry all sorts of voice, video, or information content, with social media enabling user-generated content (Web 2.0). Based on existing communication platforms like the Internet but not limited to it, the IoT represents the next step towards digitisation where all objects and people are interconnected through communication networks, in and across private, public and industrial spaces, and report about their status and/or about the status of the surrounding environment.

2.2 The IoT Is the New Age

The IoT can thus be defined as a new era of ubiquitous connectivity and intelligence, where a set of components, products, services and platforms

5

connects, virtualises and integrates everything in a communication network for digital processing.

Although the IoT is based on various disciplines and technologies like e.g. sensors, embedded systems, various communications technologies, semantic and security technologies to name but a few, it requires a specific configuration for object identification and search, open/closed data sharing, lightweight communication protocols, trade-off between local and networked based information processing, and backend integration. It also requires specific considerations of data security (e.g. location-based profiling), liability (many service providers involved) and trust ("disappearing objects").

However, the IoT will not develop without cross-cutting approaches. Focusing on vertical applications risk reinforcing silos and prevents innovation across areas. Only through the horizontal support and real-time awareness of the IoT can more powerful and disruptive innovation be delivered, and the corresponding benefits for these application areas fully leveraged. IoT promises to bring smart devices everywhere across boundaries, from the fridge to the car, from the home to the hospital to the city. Connected devices will be

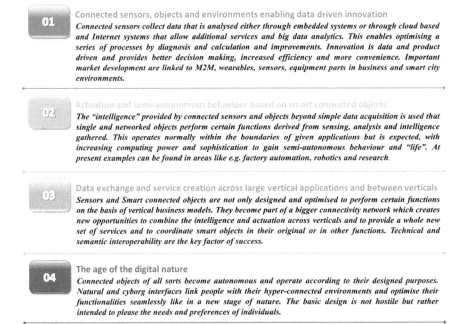

01 Connected sensors, objects and environments enabling data driven innovation
Connected sensors collect data that is analysed either through embedded systems or through cloud based and Internet systems that allow additional services and big data analytics. This enables optimising a series of processes by diagnosis and calculation and improvements. Innovation is data and product driven and provides better decision making, increased efficiency and more convenience. Important market development are linked to M2M, wearables, sensors, equipment parts in business and smart city environments.

02 Actuation and semi-autonomous behaviour based on smart connected objects
The "intelligence" provided by connected sensors and objects beyond simple data acquisition is used that single and networked objects perform certain functions derived from sensing, analysis and intelligence gathered. This operates normally within the boundaries of given applications but is expected, with increasing computing power and sophistication to gain semi-autonomous behaviour and "life". At present examples can be found in areas like e.g. factory automation, robotics and research.

03 Data exchange and service creation across large vertical applications and between verticals
Sensors and Smart connected objects are not only designed and optimised to perform certain functions on the basis of vertical business models. They become part of a bigger connectivity network which creates new opportunities to combine the intelligence and actuation across verticals and to provide a whole new set of services and to coordinate smart objects in their original or in other functions. Technical and semantic interoperability are the key factor of success.

04 The age of the digital nature
Connected objects of all sorts become autonomous and operate according to their designed purposes. Natural and cyborg interfaces link people with their hyper-connected environments and optimise their functionalities seamlessly like in a new stage of nature. The basic design is not hostile but rather intended to please the needs and preferences of individuals.

Figure 2.1   Different sequential and parallel pathways towards the Internet of Things.

powered by intelligence (embedded or in the network) to deliver new services and applications that cut across verticals.

In short, the status quo is not enough. The aim should be that the whole economy and society adopt the IoT, like what happened for mobile communication, so that it can generate maximum benefits: i) addressing societal challenges (ex: environmental protection, resource optimization, security, ageing, inclusion); ii) industrial leadership in the ICT field through new IoT ecosystems and iii) growth, employment and innovation.

2.3 The IoT Can Unleash a New Industrial and Innovation Era

IoT makes a significant reshaping of industry structures possible, with borders between products and services as well as borders between industrial sectors becoming much more blurred than today. This may materialise through:

- **Service enhanced products**: a typical example would be a car, augmented by several hundreds of embedded sensors. With such a capacity, a car becomes the focal point of an entire ecosystem that may include remote maintenance, insurance, or geolocation services. This model is similar to the iphone model, which corresponds to a product (the terminal) whose attraction and market value is significantly enhanced by the set of services it gives access to (the app store).
- **Increased efficiency and transformation in processes** – ("smart manufacturing"): the IoT makes it possible to track and integrate all production and distribution steps in the value and logistics chain and to reduce waste, increase timeliness, coordination and automation. This can vastly increase efficiency while facilitating more flexible and tailored/personalised production. For instance, supermarkets could be able to provide a complete history of each product they sell in their shelves-room, thus guaranteeing quality and offering services on top (ex: responsible farming). Factories of the future could be fully connected and automated and deliveries, including through drones and other self-driving vehicles, optimised and personalised.
- **Tighter relation supplier/buyer**: Smart, connected products expand opportunities for product differentiation, moving competition away from price alone. Knowing how customers actually use their products enhances a company's ability to segment customers, customise products, set prices to better capture value, and extend personalised value-added

services. Through capturing rich historical and product-usage data, buyers' costs of switching to a new supplier may increase. The deeper relationship with the customer hence serves to improve differentiation with them while improving its offer towards other aircraft manufacturers.

- **Increased buyer power** by giving buyers a better understanding of true product performance, allowing them to play one provider against another. Having access to product usage data can decrease their reliance on the provider for advice and support. Finally, compared with ownership models, "product as a service" business models or product-sharing services can increase buyers' power by reducing the cost of switching to a new provider.

- **New business models** enabled by smart, connected products can create a substitute for product ownership. Product-as-a-service business models, for example, allow users to have full access to a product but pay only for the amount of product they use. A variation of product-as-a-service is the shared-usage model. Companies like UBER or blablacar are examples that provide alternatives to car ownership. Equivalent substitutes for car ownership and has led traditional automakers to enter the car-sharing market with offerings such DriveNow from BMW, or Dash from Toyota.

- **New innovative actors and start-ups**: developments like the "maker culture", an extension of the DIY culture stress new and unique applications of technologies and encourage invention and prototyping, having a strong focus on using and learning practical skills and applying them creatively. SMEs can take advantage of the availability of IoT open platforms and test-beds and open source hardware and software to reduce development costs and time-to-market, and to support collaboration among businesses of different areas such as software, sensors, devices, and user businesses.

2.4 Issues to Be Tackled

Although the horizontal character of the IoT is recognized the creation of IoT ecosystems is a pre-requisite for the development of innovation and take up in the EU, which is still in an emerging phase. The IoT requires alliances between multiple sectors and stakeholders to cover an increasingly complex value chain. It also requires open platforms that can integrate many different types of equipment and application.

Another important roadblock to build IoT ecosystems relates to the lack of employee skills/knowledge, reported as being an important obstacle facing organizations in using IoT. To quote a leading medical device company, "Our sales force has been used to selling equipment, but now they need to sell IT solutions. They need to be able to convince customers on the value received by connecting their equipment".

Moreover, the IoT needs to be developed as an integral part of the Digital Single Market with a focus on creating an enabling environment for these technologies to be rolled out quickly and across the whole of Europe so as to reap economies of scale and productivity gains for our economies. This includes considering provisions to remove regulatory obstacles that prevent take up on a continental basis. In this context the European Union is willing to examine solutions to promote innovation and create a legal framework that encourages deployment.

The development of IoT may also raise privacy concerns since smart objects will collect more and new kinds of data, including personal data, and will exchange data automatically, which may lead to a perception of loss of control by citizens. IoT may further provoke ethical questions pertaining in particular to individuals' autonomy, accountability for object behaviour, or the precautionary principle. Recent examples of hacking objects have shown that the development of IoT and its integration in systems enabling key economic and societal activities may raise security and resilience issues which may require further organizational measures.

Liability is also seen as an important issue to address, in situation where wrong decisions may be taken by smart devices and connected systems. These issues are critical to acceptability of the technology by citizens. Education is needed as well as legal guidance for proper deployment conditions to make sure that the IoT serves EU values and benefits citizens genuinely, and to avoid the perception that IoT could lead to a dehumanised society controlled by the machines and/or a reinforcing of the digital divide and of social exclusion. The EU level is particularly relevant to guarantee adherence to European values such as fundamental rights, protection of integrity, inclusion, as well as openness, fair competition and open innovation.

Finally, there is a need to move into testing and deployment of IoT technologies in real-life settings. Uncertainty about business models and uncertainty about standards is generating information asymmetries and market failures preventing investment and risk-taking. In this perspective Large Scale Pilots would support testing the deployment of large amounts of sensors, or the interoperability of applications in different areas. Large Scale Pilots could also be used to investigate acceptability by users and business models. This could play an important role to address security and trust issues in an integrated manner and could contribute to certification and validation in the IoT area, as well as to certification.

2.5 Building IoT Innovation Ecosystems

IoT could become the innovation engine "par excellence", and will bring to the market entire new classes of new devices, around which sustainable innovation could take shape. Innovation in this respect can be seen from different perspectives: i) open platforms, as outlined above, can be leveraged by innovators to create new products and services, possibly in partnership with larger players; ii) for small start-up players, it is important to benefit from an innovation ecosystem where new ideas can be nurtured and incubated, before being introduced to the market.

The creation of IoT innovation ecosystems is an opportunity for Europe. Although there is no single definition for ecosystems, it is certainly important to note that they coevolve their capabilities and roles, and tend to align themselves with the directions set by one or more central companies. Leadership roles may change over time, but the function of ecosystem leader is valued by the community because it enables members to move towards shared visions to align their investments, and to find mutually supportive roles.[1] It also means that companies need to become proactive in developing mutually beneficial ("symbiotic") relationships with customers, suppliers, and even competitors.

IoT innovation ecosystems could be created around specific solutions (ex: car, home, city, hospital, devices), and be based on open platforms to deliver for instance applications and services dedicated to families of connected devices. In this context a proliferation of IoT applications and services has to lend itself on a reliable and interoperable infrastructure for device communication, smart cooperation and edge intelligence. In addition, hardware developments and new IoT products could be developed around Fablabs and IoT factories, providing all the necessary support and infrastructure to develop connected objects.

---

[1]Moore, James F. (1993). "Predators and prey: A new ecology of competition". Harvard Business Review (May/June): 75–86.

2.6 IoT Large Scale Pilots for Testing and Deployment

The deployment of IoT concerns complex systems and potentially addresses a large population of actors with different cultures and interests. Putting them together to realise a system that can operate at large scale under multiple operational constraints is still risky, and business models across complex value chains are not always well understood. The challenge is to foster the deployment of IoT solutions in Europe through integration of advanced IoT technologies across the value chain, demonstration of multiple IoT applications at scale and in a usage context, and as close as possible to operational conditions.

To move forward, the idea of deploying large scale pilots is gaining momentum globally. These pilots are designed not only to validate technological approaches from a scalability and operational perspective, but also to validate usability and user "positive reaction" to new service. From a public policy perspective, these pilots need to be driven by considerations of openness that lock-in situations and limited interoperability are avoided whilst the possibility to build open innovation on top is maximised.

Considering the important investments on IoT technologies which have already been taken at EU and Member States levels, it is evident to realize the next big step towards implementation of large scale pilots. Under Horizon 2020, the European Commission will launch a series of large scale pilots in promising domains cutting across the interest of multiple usage sectors, and cutting across different industrial sectors, both from supply and demand side perspectives. These use cases will be supported by open platforms. The pilots will not be designed as a pure technology exercise but in a way to deliver best practices in terms of technology and standards applicability, privacy and security, business models, and user acceptance. The pilots should also be used to derive methodologies to design Privacy and Security impact assessments in the IoT context.

The piloting activities will be complemented with support actions addressing challenges critically important for the take-up of IoT at the anticipated scale. These include ethics and privacy, trust and security, standards and interoperability, user acceptability, liability and sustainability, and new ways of creativity including the combination of ICT and Art. In addition the pilots will be complemented through international cooperation and specific IoT research and innovation efforts for ensuring the longer-term evolution of Internet of Things.

2.7 Alliance for Internet of Things Innovation

In the past months it became obvious that no thorough and wide ranging innovation with happen without cooperation. In order to deliver comprehensive solutions, cooperation even with potential competitors or with new partners entering the field of IoT is pivotal for two reasons: 1) one single entity cannot provide all components of a solution, and 2) because of multiple possible technical combinations and implementations, co-development reduces the risk of failure and sub-optimal solutions and provides best practices.

In order to support this process the Commission facilitated the creation of a new Alliance named AIOTI – Alliance for Internet of Things Innovation, comprising in particular industry representatives from larger but also younger IoT innovators. This Alliance, which is open by nature, and their members strive together that Europe will have the most dynamic and agile IoT ecosystem and industry in the world, with the ultimate goal to transform people's lives, drive growth, create employment and address societal challenges.

Alliance for Internet of Things Innovation - AIOTI

The AIOTI Momentum Declaration

Europe will have the most dynamic, agile IoT ecosystem and industry in the world which transforms people's lives, drives growth, creates employment and addresses societal challenges.

Today we agree, in partnership with the European Commission, that collaborative and innovation driven activities are necessary in order to drive a successful take-up of the Internet of Things.

By understanding the potential of connected things, their intelligence and smart data, we all support the creation of an IoT ecosystem, which supports openness, value creation, scalability, sustainability and co-existence.

Core principles are to cooperate and share knowledge with existing and new partners of all sizes along value chains, to adopt agile approaches and to search flexible agreements for convergence, interoperability and standardisation.

Through common reference models and IoT Large Scale Pilot activities we aim to bring IoT forward and to stimulate service creation, acceptance and take-up from the user and creator perspectives.

4th February, 2015

Figure 2.2   The Alliance Momentum declaration.

The Alliance for Internet of Things Innovation (AIOTI) is also an important tool for supporting the policy and dialogue within the Internet of Things world and within the European Commission. It builds on the work of the IoT European Research Cluster (IERC) and expands activities towards innovation within and across industries. In light of the IoT Large Scale Pilots to be funded under the Horizon 2020 Research and Innovation Program, the Alliance allows all potential stakeholders to pre-structure potential approaches in the areas of but not limited to smart living environments, smart farming, wearables, smart cities, mobility and smart environment.

Not limited to IoT Large Scale Pilots as such, the Alliance has also set up workgroups in the fields of Innovation Ecosystems, IoT Standardisation and Policy issues (trust, security, liability, privacy). Overall the alliance will help to create the necessary links and to forge cross-sectorial synergies.

2.8 Conclusions

The Internet of Things has entered the next stage and reached early adopters and the market. Yet a sound effort is necessary for providing interoperable and trustful IoT implementations. From emerging IoT Ecosystems towards IoT Large Scale Pilots, the European Commission attributes a great importance to IoT activities driven by end-user and citizen, and involving existing and new communities at an early stage.

It would be a strategic mistake not to take up the challenge for the EU to become one of the global leaders in the IoT field – Europe has today a unique opportunity to use the IoT to rejuvenate its industry, deal with its ageing population and transform its cities into places to be.

3

Internet of Things beyond the Hype: Research, Innovation and Deployment

Ovidiu Vermesan[1], Peter Friess[2], Patrick Guillemin[3], Raffaele Giaffreda[4], Hanne Grindvoll[1], Markus Eisenhauer[5], Martin Serrano[6], Klaus Moessner[7], Maurizio Spirito[8], Lars-Cyril Blystad[1] and Elias Z. Tragos[9]

[1]SINTEF, Norway
[2]European Commission, Belgium
[3]ETSI, France
[4]CREATE-NET, Italy
[5]Fraunhofer FIT, Germany
[6]National University of Ireland Galway, Ireland
[7]University of Surrey, UK
[8]ISMB, Italy
[9]FORTH, Greece

"There's a way to do it better. Find it." Thomas Edison

3.1 Internet of Things Vision

Internet of Things (IoT) is a concept and a paradigm that considers pervasive presence in the environment of a variety of things/objects that through wireless and wired connections and unique addressing schemes are able to interact with each other and cooperate with other things/objects to create new applications/services and reach common goals. In this context the research and development challenges to create a smart world are enormous. A world where the real, digital and the virtual are converging to create smart environments that make energy, transport, cities and many other areas more intelligent. The goal of the Internet of Things is to enable things to be connected anytime, anyplace, with anything and anyone ideally using any path/network and any service. Internet of Things is a new revolution of the Internet. Objects

15

make themselves recognizable and they obtain intelligence by making or enabling context related decisions thanks to the fact that they can communicate information about themselves and they can access information that has been aggregated by other things, or they can be components of complex services [71].

The various layers of the IoT value chain cover several distinct product or service categories. Sensors provide much of the data gathering, actuators act, radios/communications chips provide the underlying connectivity, micro-controllers provide the processing of that data, modules combine the radio, sensor and microcontroller, combine it with storage, and make it "insertable" into a device. Platform software provides the underlying management and billing capabilities of an IoT network, while application software presents all the information gathered in a usable and analysable format for end users. The underlying telecom infrastructure (usually wireless spectrum) provides the means of transporting the data while a service infrastructure needs to be created for the tasks of designing, installing, monitoring and servicing the IoT deployment. Companies will compete at one layer of the IoT value chain, while many will create solutions from multiple layers and functionally compete in a more vertically integrated fashion. [42].

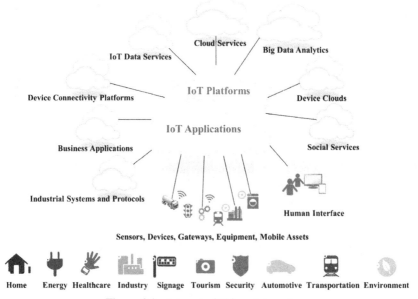

Figure 3.1 Internet of Things Integration.

The Internet of Things makes use of synergies that are generated by the convergence of Consumer, Business and Industrial Internet. The convergence creates the open, global network connecting people, data, and things. This convergence leverages the cloud to connect intelligent things that sense and transmit a broad array of data, helping creating services that would not be obvious without this level of connectivity and analytical intelligence. The use of platforms is being driven by transformative technologies such as cloud, things, and mobile. The Internet of Things and Services makes it possible to create networks incorporating the entire manufacturing process that convert factories into a smart environment. The cloud enables a global infrastructure to generate new services, allowing anyone to create content and applications for global users. Networks of things connect things globally and maintain their identity online. Mobile networks allow connection to this global infrastructure anytime, anywhere. The result is a globally accessible network of things, users, and consumers, who are available to create businesses, contribute content, generate and purchase new services.

Platforms also rely on the power of network effects, as they allow more things, they become more valuable to the other things and to users that make use of the services generated. The success of a platform strategy for IoT can be determined by connection, attractiveness and knowledge/information/data flow.

The Alliance for Internet of Things Innovation (AIOTI) was recently initiated by the European Commission in order to develop and support the dialogue and interaction among the Internet of Things (IoT) various players. The overall goal of the establishment of the AIOTI is the creation of a dynamic European IoT ecosystem to unleash the potentials of the IoT.

The AIOTI will assist the European Commission in the preparation of future IoT research as well as innovation and standardisation policies. It is also going to play an essential role in the designing of IoT Large Scale Pilots, which will be funded by the Horizon 2020 Research and Innovation Programme. The members of AIOTI will jointly work on the creation of a dynamic European IoT ecosystem. This ecosystem is going to build on the work of the IoT Research Cluster (IERC) and spill over innovation across industries and business sectors of IoT transforming ideas to IoT solutions.

The European Commission (EC) considers that IoT will be pivotal in enabling the digital single market, through new products and services. The IoT, big data, cloud computing and their related business models will be the three most important drivers of the digital economy, and in this context it is fundamental for a fully functional single market in Europe to address aspects of ownership, access, privacy and data flow – the new production factor.

New generations of networks, IoT and cloud computing are also vectors of industrial strategy. The IoT stakeholders are creating a new ecosystem that cuts across vertical areas, in convergence between the physical and digital words. It combines connectivity, data generation, processing and analytics, with actuation and new interfaces, resulting in new products and services based on platforms and software and apps.

Internet of Things developments implies that the environments, cities, buildings, vehicles, clothing, portable devices and other objects have more and more information associated with them and/or the ability to sense, communicate, network and produce new information. In addition the network technologies have to cope with the new challenges such as very high data rates, dense crowds of users, low latency, low energy, low cost and a massive number of devices. Wireless connectivity anywhere, anytime and between every-body and every-thing (smart houses, vehicles, cities, offices etc.) is gaining momentum, rendering our daily lives easier and more efficient. This momentum will continue to rise, resulting in the need to enable wireless connections between people, machines, communities, physical things, processes, content etc. anytime, in flexible, reliable and secure ways. The air interfaces for 2G, 3G, and 4G were all designed for specific use cases with certain KPIs in mind (throughput, capacity, dropped/blocked call rates etc.). However, the emerging trend of connecting everything to the Internet (IoT and Internet of Vehicles, IoV) brings up the need to go beyond such an approach. The inclusion of the above mentioned use cases pose new challenges due to the broader range of service and device classes, ranging from IoT to short range Mobile Broadband (MBB) communications (e.g. WiFi) and from high-end smartphone to low-end sensor. Furthermore, each service type/device class has more stringent requirements than ever (e.g. air interface latency in the order of 1ms) and some of these requirements are conflicting (e.g. to support very low latencies, energy and resource efficiency may not be optimal). So, the challenge is not only to increase the user rates or the capacity (as has always been so far) but also to master the heterogeneity and the trade-off between the conflicting requirements as presented in Figure 3.2 [3].

As the Internet of Things becomes established in smart factories, both the volume and the level of detail of the corporate data generated will increase. Moreover, business models will no longer involve just one company, but will instead comprise highly dynamic networks of companies and completely new value chains. Data will be generated and transmitted autonomously by

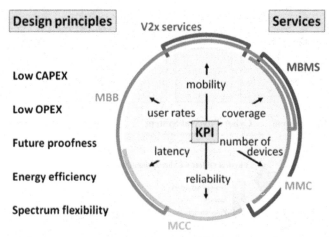

Figure 3.2 Design principles, services and related KPIs [3].

smart machines and these data will inevitably cross company boundaries. A number of specific dangers are associated with this new context – for example, data that were initially generated and exchanged in order to coordinate manufacturing and logistics activities between different companies could, if read in conjunction with other data, suddenly provide third parties with highly sensitive information about one of the partner companies that might, for example, give them an insight into its business strategies. New instruments will be required if companies wish to pursue the conventional strategy of keeping such knowledge secret in order to protect their competitive advantage. New, regulated business models will also be necessary – the raw data that are generated may contain information that is valuable to third parties and companies may therefore wish to make a charge for sharing them. Innovative business models like this will also require legal safeguards (predominantly in the shape of contracts) in order to ensure that the value added created is shared out fairly, e.g. through the use of dynamic pricing models [56].

3.1.1 Internet of Things Common Definition

The IoT is a key enabling technology for digital businesses. Approximately 3.9 billion connected things were in use in 2014 and this figure is expected to rise to 25 billion by 2020. Gartner's top 10 strategy technology trends [55] cover three themes: the merging of the real and virtual worlds, the advent of intelligence everywhere, and the technology impact of the digital business shift.

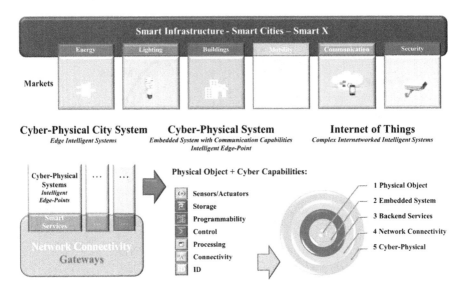

Figure 3.3   Cyber-physical sytems as building blocks of IoT applications.

The traditional distinction between network and device is starting to blur as the functionalities of the two become indistinguishable. Shifting the focus from the IoT network to the devices costs less, scales more gracefully, and leads to immediate revenues.

The systemic nature of innovation requires the need for coordination stakeholders, systems and services in interaction-intensive environments with a permanent and seamless mix of online and real-world experiences and offerings, as the IoT will consist of countless cyber-physical systems (CPS). The overlay of virtual and physical will be enabled by layered and augmented reality interfaces for interconnected things, smartphones, wearables, industrial equipment, which will exchange continuous data via edge sensor/actuator networks and context-aware applications using ubiquitous connectivity and computing by integrating technologies such as cloud edge cloud/fog and mobile. In this context the IoT applications will have real time access to intelligence about virtual and physical processes and events by open, linked and smart data.

Gartner [54, 55] identifies that the combination of data streams and services created by digitizing everything creates four basic usage models:

- Manage
- Monetize

- Operate
- Extend.

These can be applied to people, things, information, and places, and therefore the so called "Internet of Things" will be succeeded by the "Internet of Everything."

In this context the notion of network convergence using IP is fundamental and relies on the use of a common multi-service IP network supporting a wide range of applications and services.

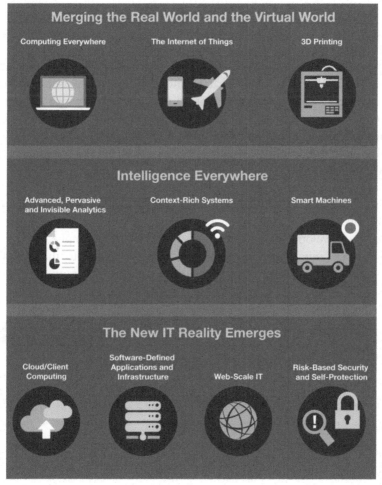

Figure 3.4 The top 10 strategic technology trends for 2015 [55].

The Internet of Things is not a single technology, it's a concept in which most new things are connected and enabled such as street lights being networked and things like embedded sensors, image recognition functionality, augmented reality, near field communication are integrated into situational decision support, asset management and new services. These bring many business opportunities and add to the complexity of IT [52].

To accommodate the diversity of the IoT, there is a heterogeneous mix of communication technologies, which need to be adapted in order to address the needs of IoT applications such as energy efficiency, security, and reliability. In this context, it is possible that the level of diversity will be scaled to a number a manageable connectivity technologies that address the needs of the IoT applications, are adopted by the market, they have already proved to be serviceable, supported by a strong technology alliance.

The Internet of Things provides solutions based on the integration of information technology, which refers to hardware and software used to store, retrieve, and process data and communications technology which includes electronic systems used for communication between individuals or groups. The rapid convergence of information and communications technology is taking place at three layers of technology innovation: the cloud, data and communication pipes/networks and device [44].

IoT will rearrange the tech landscape, again. IoT has key attributes that distinguish it from the "regular" Internet, as captured by the S-E-N-S-E framework presented in Figure 3.5. These attributes may tilt the direction of technology development and adoption, with significant implications for Tech companies, much like the transition from the fixed to the mobile Internet shifted the centre of gravity among the different actors in the value chain.

S-E-N-S-E	What the Internet of Things does	How it differs from the Internet
Sensing	*Leverages sensors attached to things (e.g. temperature, pressure, acceleration)*	*More data is generated by things with sensors than by people*
Efficient	*Adds intelligence to manual processes (e.g. reduce power usage on hot days)*	*Extends the Internet's productivity gains to things, not just people*
Networked	*Connects objects to the network (e.g. thermostats, vehicles, watches)*	*Some of the intelligence shifts from the cloud to the network's edge ("fog" computing)*
Specialized	*Customizes technology and process to specific verticals (e.g. healthcare, retail, oil)*	*Unlike the broad horizontal reach of PCs and smartphones, the IoT is very fragmented*
Everywhere	*Deployed pervasively (e.g. on the human body, in cars, homes, cities, factories)*	*Ubiquitous presence, resulting in an order of magnitude more devices and even greater security concerns*

Figure 3.5   Making S-E-N-S-E of the Internet of Things (Source: Goldman Sachs Global Investment Research).

The synergy of the access and potential data exchange opens huge new possibilities for IoT applications. Already over 50% of Internet connections are between or with things.

By 2020, over 30 billion connected things, with over 200 billion with intermittent connections are forecast. Key technologies here include embedded sensors, image recognition and NFC. By 2015, in more than 70% of enterprises, a single executable will oversee all Internet connected things. This becomes the Internet of Everything [53].

As a result of this convergence, the IoT applications require that classical industries are adapting and the technology will create opportunities for new industries to emerge and to deliver enriched and new user experiences and services.

In addition, to be able to handle the sheer number of things and objects that will be connected in the IoT, cognitive technologies and contextual intelligence are crucial. This also applies for the development of context aware applications that need to be reaching to the edges of the network through smart devices that are incorporated into our everyday life.

The Internet is not only a network of computers, but it has evolved into a network of devices of all types and sizes, vehicles, smartphones, home appliances, toys, cameras, medical instruments and industrial systems, all connected, all communicating and sharing information all the time.

The Internet of Things had until recently different means at different levels of abstractions through the value chain, from lower level semiconductor through the service providers.

The Internet of Things is a "global concept" and requires a common definition. Considering the wide background and required technologies, from sensing device, communication subsystem, data aggregation and pre-processing to the object instantiation and finally service provision, generating an unambiguous definition of the "Internet of Things" is non-trivial.

The IERC is actively involved in ITU-T Study Group 13, which leads the work of the International Telecommunications Union (ITU) on standards for next generation networks (NGN) and future networks and has been part of the team which has formulated the following definition [67]: *"**Internet of things (IoT)**: A global infrastructure for the information society, enabling advanced services by interconnecting (physical and virtual) things based on existing and evolving interoperable information and communication technologies. NOTE 1 – Through the exploitation of identification, data capture, processing and communication capabilities, the IoT makes full use of things to offer services to all kinds of applications, whilst ensuring*

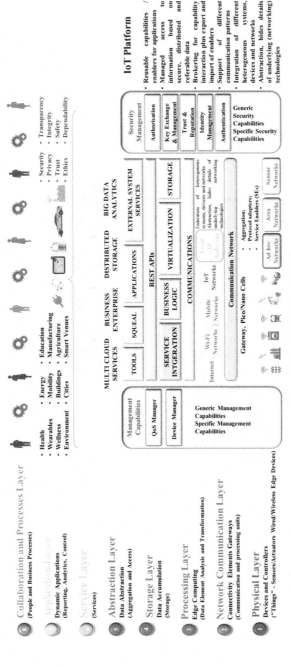

Figure 3.6 IoT Architectural View.

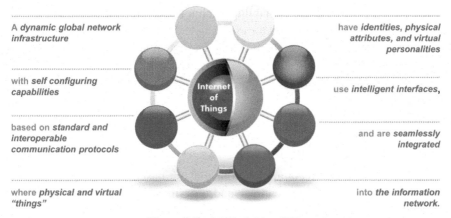

A *dynamic global network infrastructure*

with *self configuring capabilities*

based on *standard and interoperable communication protocols*

where *physical and virtual "things"*

have *identities, physical attributes, and virtual personalities*

use *intelligent interfaces,*

and are *seamlessly integrated*

into *the information network.*

Figure 3.7 IoT Definition [70].

that security and privacy requirements are fulfilled. NOTE 2 – From a broader perspective, the IoT can be perceived as a vision with technological and societal implications."

The IERC definition [70] states that IoT is *"A dynamic global network infrastructure with self-configuring capabilities based on standard and interoperable communication protocols where physical and virtual "things" have identities, physical attributes, and virtual personalities and use intelligent interfaces, and are seamlessly integrated into the information network."*.

3.2 IoT Strategic Research and Innovation Directions

The development of enabling technologies such as nanoelectronics, communications, sensors, smart phones, embedded systems, cloud networking, network virtualization and software will be essential to provide to things the capability to be connected all the time everywhere. This will also support important future IoT product innovations affecting many different industrial sectors. Some of these technologies such as embedded or cyber-physical systems form the edges of the "Internet of Things" bridging the gap between cyber space and the physical world of real "things", and are crucial in enabling the "Internet of Things" to deliver on its vision and become part of bigger systems in a world of "systems of systems".

The final report of the Key Enabling Technologies (KET), of the High-Level Expert Group [45] identified the enabling technologies, crucial to many of the existing and future value chains of the European economy:

- Nanotechnologies
- Micro and Nano electronics
- Photonics
- Biotechnology
- Advanced Materials
- Advanced Manufacturing Systems

As such, IoT creates intelligent applications that are based on the supporting KET's identified, as IoT applications address smart environments either physical or at cyber-space level, and in real time.

To this list of key enablers, we can add the global deployment of IPv6 across the World enabling a global and ubiquitous addressing of any communicating smart thing.

From a technology perspective, the continuous increase in the integration density proposed by Moore's Law was made possible by a dimensional scaling: in reducing the critical dimensions while keeping the electrical field constant, one obtained at the same time a higher speed and a reduced power consumption of a digital MOS circuit: these two parameters became driving forces of the microelectronics industry along with the integration density.

The International Technology Roadmap for Semiconductors has emphasized in its early editions the "miniaturization" and its associated benefits in terms of performances, the traditional parameters in Moore's Law. This trend for increased performances will continue, while performance can always be traded against power depending on the individual application, sustained by the incorporation into devices of new materials, and the application of new transistor concepts. This direction for further progress is labelled "More Moore".

The second trend is characterized by functional diversification of semiconductor-based devices. These non-digital functionalities do contribute to the miniaturization of electronic systems, although they do not necessarily scale at the same rate as the one that describes the development of digital functionality. Consequently, in view of added functionality, this trend may be designated "More-than-Moore" [48].

Mobile data traffic is projected to double each year between now and 2015 and mobile operators will find it increasingly difficult to provide the bandwidth requested by customers. In many countries there is no additional spectrum that can be assigned and the spectral efficiency of mobile networks is reaching its physical limits. Proposed solutions are the seamless integration of existing Wi-Fi networks into the mobile ecosystem. This will have a direct impact on Internet of Things ecosystems. The chips designed to accomplish this integration are known as "multicom" chips. Wi-Fi and baseband communications are expected to converge and the architecture of mobile devices is likely to change and the baseband chip is expected to take control of the routing so the connectivity components are connected to the baseband or integrated in a single silicon package. As a result of this architecture change, an increasing share of the integration work is likely done by baseband manufacturers (ultra -low power solutions) rather than by handset producers.

Today many European projects and initiatives address Internet of Things technologies and knowledge. Given the fact that these topics can be highly diverse and specialized, there is a strong need for integration of the individual results. Knowledge integration, in this context is conceptualized as the process through which disparate, specialized knowledge located in multiple projects across Europe is combined, applied and assimilated.

The Strategic Research and Innovation Agenda (SRIA) is the result of a discussion involving the projects and stakeholders involved in the IERC activities, which gather the major players of the European ICT landscape addressing IoT technology priorities that are crucial for the competitiveness of European industry.

IERC Strategic Research and Innovation Agenda covers the important issues and challenges for the Internet of Things technology. It provides the vision and the roadmap for coordinating and rationalizing current and future research and development efforts in this field, by addressing the different enabling technologies covered by the Internet of Things concept and paradigm.

Many other technologies are converging to support and enable IoT applications. These technologies are summarised as:

- IoT architecture
- Identification
- Communication
- Networks technology
- Network discovery
- Software and algorithms

- Hardware technology
- Data and signal processing
- Discovery and search engine
- Network management
- Power and energy storage
- Security, trust, dependability and privacy
- Interoperability
- Standardization

The Strategic Research and Innovation Agenda is developed with the support of a European-led community of interrelated projects and their stakeholders, dedicated to the innovation, creation, development and use of the Internet of Things technology.

Since the release of the first version of the Strategic Research and Innovation Agenda, we have witnessed active research on several IoT topics. On the one hand this research filled several of the gaps originally identified in the Strategic Research and Innovation Agenda, whilst on the other it created new challenges and research questions. Recent advances in areas such as cloud computing, cyber-physical systems, autonomic computing, and social networks have changed the scope of the Internet of Thing's convergence even more so. The Cluster has a goal to provide an updated document each year that records the relevant changes and illustrates emerging challenges. The updated release of this Strategic Research and Innovation Agenda builds incrementally on previous versions [70, 71, 92, 93] and highlights the main research topics that are associated with the development of IoT enabling technologies, infrastructures and applications with an outlook towards 2020 [82].

The research items introduced will pave the way for innovative applications and services that address the major economic and societal challenges underlined in the EU 2020 Digital Agenda [83].

The IERC Strategic Research and Innovation Agenda is developed incrementally based on its previous versions and focus on the new challenges being identified in the last period.

The updated release of the Strategic Research and Innovation Agenda is highlighting the main research topics that are associated with the development of IoT infrastructures and applications, with an outlook towards 2020 [82].

The timeline of the Internet of Things Strategic Research and Innovation Agenda covers the current decade with respect to research and the following years with respect to implementation of the research results. Of course, as the Internet and its current key applications show, we anticipate unexpected trends will emerge leading to unforeseen and unexpected development paths.

The Cluster has involved experts working in industry, research and academia to provide their vision on IoT research challenges, enabling technologies and the key applications, which are expected to arise from the current vision of the Internet of Things.

The IoT Strategic Research and Innovation Agenda covers in a logical manner the vision, the technological trends, the applications, the technology enablers, the research agenda, timelines, priorities, and finally summarises in two tables the future technological developments and research needs.

The field of the Internet of Things is based on the paradigm of supporting the IP protocol to all edges of the Internet and on the fact that at the edge of the network many (very) small devices are still unable to support IP protocol stacks. This means that solutions centred on minimum Internet of Things devices are considered as an additional Internet of Things paradigm *without IP to all access edges*, due to their importance for the development of the field.

3.2.1 IoT Applications and Deployment Scenarios

The IERC vision is that "the major objectives for IoT are the creation of smart environments/spaces and self-aware things (for example: smart transport, products, cities, buildings, rural areas, energy, health, living, etc.) for climate, food, energy, mobility, digital society and health applications" [70].

The outlook for the future is the emerging of a network of interconnected uniquely identifiable objects and their virtual representations in an Internet alike structure that is positioned over a network of interconnected computers allowing for the creation of a new platform for economic growth.

Smart is the new green as defined by Frost & Sullivan [49] and the green products and services will be replaced by smart products and services. Smart products have a real business case, can typically provide energy and efficiency savings of up to 30 per cent, and generally deliver a two- to three-year return on investment. This trend will help the deployment of Internet of Things applications and the creation of smart environments and spaces.

At the city level, the integration of technology and quicker data analysis will lead to a more coordinated and effective civil response to security and safety (law enforcement and blue light services); higher demand for outsourcing security capabilities.

At the building level, security technology will be integrated into systems and deliver a return on investment to the end-user through leveraging the technology in multiple applications (HR and time and attendance, customer behaviour in retail applications etc.).

There will be an increase in the development of "Smart" vehicles which have low (and possibly zero) emissions. They will also be connected to infrastructure. Additionally, auto manufacturers will adopt more use of "Smart" materials.

The key focus will be to make the city smarter by optimizing resources, feeding its inhabitants by urban farming, reducing traffic congestion, providing more services to allow for faster travel between home and various destinations, and increasing accessibility for essential services. It will become essential to have intelligent security systems to be implemented at key junctions in the city. Various types of sensors will have to be used to make this a reality. Sensors are moving from "smart" to "intelligent".

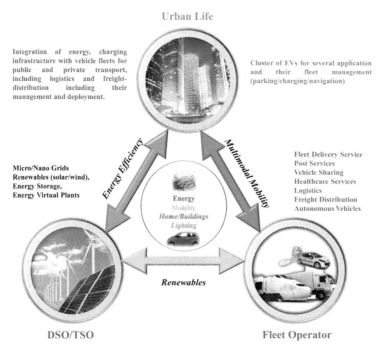

Figure 3.8   IoT applications for integration of different vertical sectors.

Wastewater treatment plants will evolve into bio-refineries. New, innovative wastewater treatment processes will enable water recovery to help close the growing gap between water supply and demand.

Self-sensing controls and devices will mark new innovations in the Building Technologies space. Customers will demand more automated, self-controlled solutions with built in fault detection and diagnostic capabilities.

Development of smart implantable chips that can monitor and report individual health status periodically will see rapid growth.

Smart pumps and smart appliances/devices are expected to be significant contributors towards efficiency improvement. Process equipment with in built "smartness" to self-assess and generate reports on their performance, enabling efficient asset management, will be adopted.

The Industrial Internet starts with embedding sensors and other advanced instrumentation in an array of machines from the simple to the highly complex. This allows the collection and analysis of an enormous amount of data, which can be used to improve machine performance, and inevitably the efficiency of the systems and networks that link them. Even the data itself can become "intelligent," instantly knowing which users it needs to reach.

Consumer IoT is essentially wireless, while the industrial IoT has to deal with an installed base of millions of devices that could potentially become part of this network (many legacy systems installed before IP deployment). These industrial objects are linked by wires that provides the reliable communications needed. The industrial IoT has to consider the legacy using specialised protocols, including Lonworks, DeviceNet, Profibus and CAN and they will be connected into this new network of networks through gateways.

The automation and management of asset-intensive enterprises will be transformed by the rise of the IoT, Industry 4.0, or simply Industrial Internet. Compared with the Internet revolution, many product and asset management solutions have laboured under high costs and poor connectivity and performance. This is now changing. New high-performance systems that can support both Internet and Cloud connectivity as well as predictive asset management are reaching the market. New cloud computing models, analytics, and aggregation technologies enable broader and low cost application of analytics across these much more transparent assets. These developments have the potential to radically transform products, channels, and company business models. This will create disruptions in the business and opportunities for all types of organizations – OEMs, technology suppliers, system integrators, and global consultancies. There may be the opportunity to overturn established business models, with a view toward answering customer pain points and also growing the market in segments that cannot be served economically with today's offerings. Mobility, local diagnostics, and remote asset monitoring are important components of these new solutions, as all market participants need ubiquitous access to their assets, applications, and customers. Real-time mobile applications support EAM, MRO, inventory management, inspections, workforce management, shop floor interactions, facilities management, field service automation, fleet management, sales and marketing, machine-to-machine (M2M), and many others [57].

In this context the concept of Internet of Energy requires web based architectures to readily guarantee information delivery on demand and to change the traditional power system into a networked Smart Grid that is largely automated, by applying greater intelligence to operate, enforce policies, monitor and self-heal when necessary. This requires the integration and interfacing of the power grid to the network of data represented by the Internet, embracing energy generation, transmission, delivery, substations, distribution control, metering and billing, diagnostics, and information systems to work seamlessly and consistently.

The concept enables the ability to produce, store and efficiently use energy, while balancing the supply/demand by using a cognitive Internet of Energy that harmonizes the energy grid by processing the data, information and knowledge via the Internet. The Internet of Energy concept leverages on the information highway provided by the Internet to link devices and services with the distributed smart energy grid that is the highway for renewable energy resources allowing stakeholders to use green technologies and sell excess energy back to the utility. The concept has the energy management element in the centre of the communication and exchange of data and energy.

The Smart-X environments are implemented using CPS building blocks integrated into Internet of X applications connected through the Internet and enabling seamless and secure interactions and cooperation of intelligent embedded systems over heterogeneous communication infrastructures.

It is expected that this "development of smart entities will encourage development of the novel technologies needed to address the emerging challenges of public health, aging population, environmental protection and climate change, conservation of energy and scarce materials, enhancements to safety and security and the continuation and growth of economic prosperity." The IoT applications are further linked with Green ICT, as the IoT will drive energy-efficient

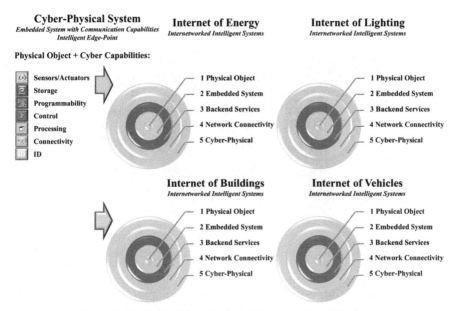

Figure 3.9 CPS building blocks for Internet of X applications.

applications such as smart grid, connected electric cars, energy-efficient buildings, thus eventually helping in building green intelligent cities.

3.3 IoT Smart-X Applications

The IoT applications are addressing the societal needs and the advancements to enabling technologies such as nanoelectronics and cyber-physical systems continue to be challenged by a variety of technical (i.e., scientific and engineering), institutional, and economical issues.

The list is focusing to the applications chosen by the IERC as priorities for the next years and it provides the research challenges for these applications. While the applications themselves might be different, the research challenges are often the same or similar.

3.3.1 Wearables

Wearables are integrating key technologies (e.g. nanoelectronics, organic electronics, sensing, actuating, communication, low power computing, visualisation and embedded software) into intelligent systems to bring new functionalities into clothes, fabrics, patches, watches and other body-mounted devices.

Figure 3.10   Smart wristbands and watches – connected IoT devices.

These intelligent edge devices are more and more part of integrated IoT solutions and assist humans in monitoring, situational awareness and decision making. They can provide actuating functions for fully automated closed-loop solutions that are used in healthcare, well-being, safety, security, infotainment applications and connected with smart buildings, energy, lighting, mobility or smart cities IoT applications. With more than 35 million connected wearable devices in use by the end of 2014, developers are pushing the technological integration into IoT applications looking for the innovation opportunities in different domains. Today, Over 75% of consumers with wearable devices stop using them within 6 months. The challenge for developers is to leverage actionable data to create apps that are seamlessly integrated into everyday life and integrate them with other IoT applications.

Creating a seamless user experience is essential for wearable application success. Leveraging tools to implement gesture-centric interfaces will allow users to make the most of limited surfaces of the wearables. The integration into common IoT platforms where developers can access data gathered from wearable devices is essential recombining datasets to develop applications for specific use cases. The industrial sector offers many opportunities for developers with the augmented reality headsets needed to be used to integrate wearables for solving real problems in the industrial sector.

The market for wearable computing is expected to grow six-fold, from 46 million units in 2014 to 285 million units in 2018 [51]. Wearable computing applications include everything from fitness trackers, health monitors, smart watches that provide new ways to interact with and utilize your smartphone, to augmented reality glasses wearable computing device.

Fitness tracking is the biggest application today and this opens the opportunities for watches that are capable of tracking blood pressure, glucose, temperature, pulse rate and other vital parameters measured every few seconds for a long period of time to be integrated in new kinds of healthcare applications. Glasses for augmented reality can be another future wearable application.

3.3.2 Smart Health, Wellness and Ageing Well

The market for health monitoring devices is currently characterised by application-specific solutions that are mutually non-interoperable and are made up of diverse architectures. While individual products are designed to cost targets, the long-term goal of achieving lower technology costs across current and future sectors will inevitably be very challenging unless a more coherent approach is used. The IoT can be used in clinical care where hospitalized patients whose physiological status requires close attention can be constantly monitored using IoT -driven, non-invasive monitoring. This requires sensors to collect comprehensive physiological information and uses gateways and the cloud to analyse and store the information and then send the analysed data wirelessly to caregivers for further analysis and review. These techniques improve the quality of care through constant attention and lower the cost of care by eliminating the need for a caregiver to actively engage in data collection and analysis. In addition the technology can be used for remote monitoring using small, wireless solutions connected through the IoT. These solutions can be used to securely capture patient health data from a variety of sensors, apply complex algorithms to analyse the data and then share it through wireless connectivity with medical professionals who can make appropriate health recommendations.

The links between the many applications in health monitoring are:

- Applications require the gathering of data from sensors.
- Applications must support user interfaces and displays.
- Applications require network connectivity for access to infrastructural services.
- Applications have in-use requirements such as low power, robustness, durability, accuracy and reliability.

IoT applications are pushing the development of platforms for implementing ambient assisted living (AAL) systems that will offer services in the areas of assistance to carry out daily activities, health and activity monitoring, enhancing safety and security, getting access to medical and emergency systems, and facilitating rapid health support.

The main objective is to enhance life quality for people who need permanent support or monitoring, to decrease barriers for monitoring important health parameters, to avoid unnecessary healthcare costs and efforts, and to provide the right medical support at the right time.

The IoT plays an important role in healthcare applications, from managing chronic diseases at one end of the spectrum to preventing disease at the other.

Challenges exist in the overall cyber-physical infrastructure (e.g., hardware, connectivity, software development and communications), specialized processes at the intersection of control and sensing, sensor fusion and decision making, security, and the compositionality of cyber-physical systems. Proprietary medical devices in general were not designed for interoperation with other medical devices or computational systems, necessitating advancements in networking and distributed communication within cyber-physical architectures. Interoperability and closed loop systems appears to be the key for success. System security will be critical as communication of individual patient data is communicated over cyber-physical networks. In addition, validating data acquired from patients using new cyber-physical technologies against existing gold standard data acquisition methods will be a challenge. Cyber-physical technologies will also need to be designed to operate with minimal patient training or cooperation [91].

New and innovative technologies are needed to cope with the trends on wired, wireless, high-speed interfaces, miniaturization and modular design approaches for products having multiple technologies integrated.

IoT applications have a market potential for electronic health services and connected telecommunication industry with the possibility of building ecosystems in different application areas. Medical expenditures are in the range of 10% of the European gross domestic product. The market segment of telemedicine, one of lead markets of the future will have growth rates of more than 19%.

The smart living environments at home, at work, in public spaces should be based upon integrated systems of a range of IoT-based technologies and services with user-friendly configuration and management of connected technologies for indoors and outdoors.

These systems can provide seamless services and handle flexible connectivity while users are switching contexts and moving in their living

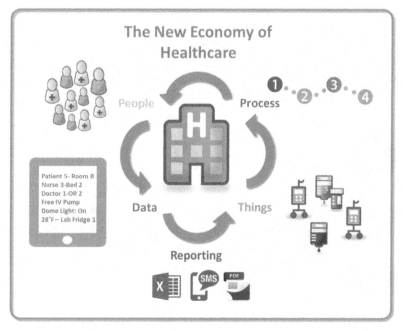

Figure 3.11 Internet of Everything and the new economy of healthcare [81].

environments and be integrated with other application domains such as energy, transport, or smart cities. The advanced IoT technologies, using and extending available open service platforms, standardised ontologies and open standardised APIs can offer many of such smart environment developments.

These IoT technologies can propose user-centric multi-disciplinary solutions that take into account the specific requirements for accessibility, usability, cost efficiency, personalisation and adaptation arising from the application requirements.

3.3.3 Smart Homes and Buildings

The rise of Wi-Fi's role in home automation has primarily come about due to the networked nature of deployed electronics where electronic devices (TVs and AV receivers, mobile devices, etc.) have started becoming part of the home IP network and due the increasing rate of adoption of mobile computing devices (smartphones, tablets, etc.).

Several organizations are working to equip homes with technology that enables the occupants to use a single device to control all electronic devices

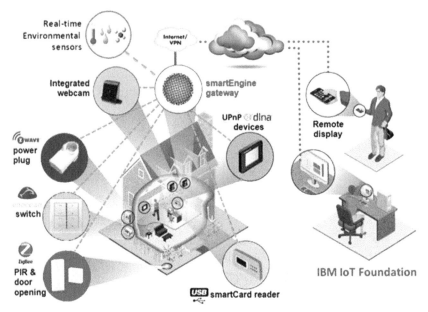

Figure 3.12   Home equipment and appliances [78].

and appliances. The solutions focus primarily on environmental monitoring, energy management, assisted living, comfort, and convenience. The solutions are based on open platforms that employ a network of intelligent sensors to provide information about the state of the home. These sensors monitor systems such as energy generation and metering; heating, ventilation, and air conditioning (HVAC); lighting; security; and environmental key performance indicators. The information is processed and made available through a number of access methods such as touch screens, mobile phones, and 3–D browsers [117]. The networking aspects are bringing online streaming services or network playback, while becoming a mean to control of the device functionality over the network. At the same time mobile devices ensure that consumers have access to a portable 'controller' for the electronics connected to the network. Both types of devices can be used as gateways for IoT applications. In this context many companies are considering building platforms that integrate the building automation with entertainment, healthcare monitoring, energy monitoring and wireless sensor monitoring in the home and building environments.

IoT applications using sensors to collect information about operating conditions combined with cloud hosted analytics software that analyse disparate data points will help facility managers become far more proactive about managing buildings at peak efficiency.

Issues of building ownership (i.e., building owner, manager, or occupants) challenge integration with questions such as who pays initial system cost and who collects the benefits over time. A lack of collaboration between the subsectors of the building industry slows new technology adoption and can prevent new buildings from achieving energy, economic and environmental performance targets.

Integration of cyber physical systems both within the building and with external entities, such as the electrical grid, will require stakeholder cooperation to achieve true interoperability. As in all sectors, maintaining security will be a critical challenge to overcome [91].

Within this field of research the exploitation of the potential of wireless sensor networks (WSNs) to facilitate intelligent energy management in buildings, which increases occupant comfort while reducing energy demand, is highly relevant. In addition to the obvious economic and environmental gains from the introduction of such intelligent energy management in buildings other positive effects will be achieved. Not least of which is the simplification of building control; as placing monitoring, information feedback equipment and control capabilities in a single location will make a buildings' energy management system easier to handle for the building owners, building managers, maintenance crews and other users of the building.

Using the Internet together with energy management systems also offers an opportunity to access a buildings' energy information and control systems from a laptop or a Smartphone placed anywhere in the world. This has a huge potential for providing the managers, owners and inhabitants of buildings with energy consumption feedback and the ability to act on that information.

The perceived evolution of building system architectures includes an adaptation level that will dynamically feed the automation level with control logic, i.e. rules. Further, in the IoT approach, the management level has also to be made available transversally as configuration; discovery and monitoring services must be made accessible to all levels. Algorithms and rules have also to be considered as Web resources in a similar way as for sensors and actuators. The repartition of roles for a classical building automation system to the new web of things enabled architecture is different and in this context, future works will have to be carried on to find solutions to minimize the transfer of data and the distribution of algorithms [46].

In the context of the future 'Internet of Things', Intelligent Building Management Systems can be considered part of a much larger information system. This system is used by facilities managers in buildings to manage energy use and energy procurement and to maintain buildings systems. It is based on the infrastructure of the existing Intranets and the Internet, and therefore utilises the same standards as other IT devices. Within this context reductions in the cost and reliability of WSNs are transforming building automation, by making the maintenance of energy efficient healthy productive work spaces in buildings increasingly cost effective [80].

3.3.4 Smart Energy

There is increasing public awareness about the changing paradigm of our policy in energy supply, consumption and infrastructure. For several reasons our future energy supply should no longer be based on fossil resources. Neither is nuclear energy a future proof option. In consequence future energy supply needs to be based largely on various renewable resources. Increasingly focus must be directed to our energy consumption behaviour. Because of its volatile nature such supply demands an intelligent and flexible electrical grid which is able to react to power fluctuations by controlling electrical energy sources (generation, storage) and sinks (load, storage) and by suitable reconfiguration. Such functions will be based on networked intelligent devices (appliances, micro-generation equipment, infrastructure, consumer products) and grid infrastructure elements, largely based on IoT concepts. Although this ideally requires insight into the instantaneous energy consumption of individual loads (e.g. devices, appliances or industrial equipment) information about energy usage on a per-customer level is a suitable first approach.

Future energy grids are characterized by a high number of distributed small and medium sized energy sources and power plants which may be combined virtually ad hoc to virtual power plants; moreover in the case of energy outages or disasters certain areas may be isolated from the grid and supplied from within by internal energy sources such as photovoltaics on the roofs, block heat and power plants or energy storages of a residential area ("islanding").

A grand challenge for enabling technologies such as cyber-physical systems is the design and deployment of an energy system infrastructure that is able to provide blackout free electricity generation and distribution, is flexible enough to allow heterogeneous energy supply to or withdrawal from the grid, and is impervious to accidental or intentional manipulations. Integration of cyber-physical systems engineering and technology to the existing electric grid and other utility systems is a challenge. The increased system complexity

Figure 3.13 Smart Energy Concept [75].

poses technical challenges that must be considered as the system is operated in ways that were not intended when the infrastructure was originally built. As technologies and systems are incorporated, security remains a paramount concern to lower system vulnerability and protect stakeholder data [91]. These challenges will need to be address as well by the IoT applications that integrate heterogeneous cyber-physical systems.

The developing Smart Grid is expected to implement a new concept of transmission network which is able to efficiently route the energy which is produced from both concentrated and distributed plants to the final user with high security and quality of supply standards. Therefore the Smart Grid is expected to be the implementation of a kind of "Internet" in which the energy packet is managed similarly to the data packet – across routers and gateways which autonomously can decide the best pathway for the packet to reach its destination with the best integrity levels. In this respect the "Internet of Energy" concept is defined as a network infrastructure based on standard and interoperable communication transceivers, gateways and protocols that will allow a real time balance between the local and the global generation and storage capability with the energy demand. This will also allow a high level of consumer awareness and involvement.

The Internet of Energy (IoE) provides an innovative concept for power distribution, energy storage, grid monitoring and communication. It will allow units of energy to be transferred when and where it is needed. Power consumption monitoring will be performed on all levels, from local individual devices up to national and international level [110]. In the long run electro mobility will become another important element of smart power grids. Electric vehicles (EVs) might act as a power load as well as moveable energy storage linked as IoT elements to the energy information grid (smart grid). IoT enabled smart grid control may need to consider energy demand and offerings in the residential areas and along the major roads based on traffic forecast. EVs will be able to act as sink or source of energy based on their charge status, usage schedule and energy price which again may depend on abundance of (renewable) energy in the grid. This is the touch point from where the following telematics IoT scenarios will merge with smart grid IoT.

Latencies are critical when talking about electrical control loops. Even though not being a critical feature, low energy dissipation should be mandatory. In order to facilitate interaction between different vendors' products the technology should be based on a standardized communication protocol stack. When dealing with a critical part of the public infrastructure, data security is of the highest importance. In order to satisfy the extremely high requirements on reliability of energy grids, the components as well as their interaction must feature the highest reliability performance.

Many IoT applications will go beyond one industrial sector. Energy, mobility and home/buildings sectors will share data through energy gateways that will control the transfer of energy and information.

Sophisticated and flexible data filtering, data mining and processing procedures and systems will become necessary in order to handle the high amount of raw data provided by billions of data sources. System and data models need to support the design of flexible systems which guarantee a reliable and secure real-time operation.

3.3.5 Smart Mobility and Transport

The connection of vehicles to the Internet gives rise to a wealth of new possibilities and applications which bring new functionalities to the individuals and/or the making of transport easier and safer. In this context the concept of Internet of Vehicles (IoV) [110] connected with the concept of Internet of Energy (IoE) represent future trends for smart transportation and mobility applications.

At the same time creating new mobile ecosystems based on trust, security and convenience to mobile/contactless services and transportation applications will ensure security, mobility and convenience to consumer-centric transactions and services.

Representing human behaviour in the design, development, and operation of cyber physical systems in autonomous vehicles is a challenge. Incorporating human-in-the-loop considerations is critical to safety, dependability, and predictability. There is currently limited understanding of how driver behaviour will be affected by adaptive traffic control cyber physical systems. In addition, it is difficult to account for the stochastic effects of the human driver in a mixed traffic environment (i.e., human and autonomous vehicle drivers) such as that found in traffic control cyber physical systems. Increasing integration calls for security measures that are not physical, but more logical while still ensuring there will be no security compromise. As cyber physical systems become more complex and interactions between components increases, safety and security will continue to be of paramount importance [91]. All these elements are of the paramount importance for the IoT ecosystems developed based on these enabling technologies.

Self-driving vehicles today are in the prototype phase and the idea is becoming just another technology on the computing industry's parts list. By using automotive vision chips that can be used to help vehicles understand the environment around them by detecting pedestrians, traffic lights, collisions, drowsy drivers, and road lane markings. Those tasks initially are more the sort of thing that would help a driver in unusual circumstances rather than take over full time. But they're a significant step in the gradual shift toward the computer-controlled vehicles that Google, Volvo, and other companies are working on [88]. The image below shows a footage of what the on-board Google Car's computer "sees" and how it detects other vehicles, pedestrians, and traffic lights [86].

These scenarios are, not independent from each other and show their full potential when combined and used for different applications.

Technical elements of such systems are smart phones and smart vehicle on-board units which acquire information from the user (e.g. position, destination and schedule) and from on board systems (e.g. vehicle status, position, energy usage profile, driving profile). They interact with external systems (e.g. traffic control systems, parking management, vehicle sharing managements, electric vehicle charging infrastructure). Moreover they need to initiate and perform the related payment procedures.

The concept of Internet of Vehicles (IoV) is the next step for future smart transportation and mobility applications and requires creating new mobile ecosystems based on trust, security and convenience to mobile/contactless services and transportation applications in order to ensure security, mobility and convenience to consumer-centric transactions and services.

Figure 3.14   Google vehicle vision [86].

Smart sensors in the road and traffic control infrastructures need to collect information about road and traffic status, weather conditions, etc. This requires robust sensors (and actuators) which are able to reliably deliver information to the systems mentioned above. Such reliable communication needs to be based on M2M communication protocols which consider the timing, safety, and security constraints. The expected high amount of data will require sophisticated data mining strategies. Overall optimisation of traffic flow and energy usage may be achieved by collective organisation among the individual vehicles.

When dealing with information related to individuals' positions, destinations, schedules, and user habits, privacy concerns gain highest priority. They even might become road blockers for such technologies. Consequently not only secure communication paths but also procedures which guarantee anonymity and de-personalization of sensible data are of interest.

Connectivity will revolutionize the environment and economics of vehicles in the future: first through connection among vehicles and intelligent infrastructures, second through the emergence of an ecosystem of services around smarter and more autonomous vehicles.

In this context the successful deployment of safe and autonomous vehicles (SAE[1] international level 5, full automation) in different use case scenarios, using local and distributed information and intelligence is an important

---

[1] Society of Automotive Engineers, J3016 standard.

achievement. This is based on real-time reliable platforms managing mixed mission and safety critical vehicle services, advanced sensors/actuators, navigation and cognitive decision-making technology, interconnectivity between vehicles (V2V) and vehicle to infrastructure (V2I) communication. There is a need to demonstrate in real life environments (i.e. highways, congested urban environment, and/or dedicated lanes), mixing autonomous connected vehicles and legacy vehicles the functionalities in order to evaluate and demonstrate dependability, robustness and resilience of the technology over longer period of time and under a large variety of conditions.

The introduction of the autonomous vehicles will enable the development of service ecosystems around vehicles and multi-modal mobility, considering that the vehicle includes multiple embedded information sources around which information services may be constructed. The information may be used for other services (i.e. maintenance, personalised insurance, vehicle behaviour monitoring and diagnostic, security and autonomous cruise, etc.).

The emergence of these services will be supported by open service platforms that communicate and exchange information with the vehicle embedded information sources and to vehicle surrounding information, with the goal of providing personalised services to drivers. Possible barriers to the deployment of autonomous vehicles and ecosystems are the robustness sensing/actuating the environment, overall user acceptance, the economic, ethical, legal and regulatory issues.

3.3.6 Smart Manufacturing and Industrial Internet of Things

The role of the Internet of Things is becoming more prominent in enabling access to devices and machines, which in manufacturing systems, were hidden in well-designed silos. This evolution will allow the IT to penetrate further the digitized manufacturing systems. The IoT will connect the factory to a whole new range of applications, which run around the production. This could range from connecting the factory to the smart grid, sharing the production facility as a service or allowing more agility and flexibility within the production systems themselves. In this sense, the production system could be considered one of the many Internets of Things (IoT), where a new ecosystem for smarter and more efficient production could be defined.

The first evolutionary step towards a shared smart factory could be demonstrated by enabling access to today's external stakeholders in order to interact with an IoT-enabled manufacturing system. These stakeholders could include the suppliers of the productions tools (e.g. machines, robots), as well as the production logistics (e.g. material flow, supply chain management), and maintenance and re-tooling actors. An IoT-based architecture that challenges the hierarchical and closed factory automation pyramid, by allowing the above-mentioned stakeholders to run their services in multiple tier flat production system is proposed in [186]. This means that the services and applications of tomorrow do not need to be defined in an intertwined and strictly linked manner to the physical system, but rather run as services in a shared physical world. The room for innovation in the application space could be increased in the same degree of magnitude as this has been the case for embedded applications or Apps, which have exploded since the arrival of smart phones (i.e. the provision of a clear and well standardized interface to the embedded hardware of a mobile phone to be accessed by all types of Apps).

Enterprises are making use of the huge amount of data available, business analytics, cloud services, enterprise mobility and many others to improve the way businesses are being conducted. These technologies include big data and business analytics software, cloud services, embedded technology, sensor networks/sensing technology, RFID, GPS, M2M, mobility, security and ID recognition technology, wireless network and standardisation.

One key enabler to this ICT-driven smart and agile manufacturing lies in the way we manage and access the physical world, where the sensors, the actuators, and also the production unit should be accessed, and managed in the same or at least similar IoT standard interfaces and technologies. These devices are then providing their services in a well-structured manner, and can be managed and orchestrated for a multitude of applications running in parallel.

The convergence of microelectronics and micromechanical parts within a sensing device, the ubiquity of communications, the rise of micro-robotics, the customization made possible by software will significantly change the world of manufacturing. In addition, broader pervasiveness of telecommunications in many environments is one of the reasons why these environments take the shape of ecosystems.

Some of the main challenges associated with the implementation of cyber-physical systems in include affordability, network integration, and the interoperability of engineering systems.

Most companies have a difficult time justifying risky, expensive, and uncertain investments for smart manufacturing across the company and factory level. Changes to the structure, organization, and culture of manufacturing occur slowly, which hinders technology integration. Pre-digital age control systems are infrequently replaced because they are still serviceable. Retrofitting these existing plants with cyber-physical systems is difficult and expensive. The lack of a standard industry approach to production management results in customized software or use of a manual approach. There is also a need for a unifying theory of non-homogeneous control and communication systems [91].

3.3.7 Smart Cities

A smart city is defined as a city that monitors and integrates conditions of all of its critical infrastructures, including roads, bridges, tunnels, rail/subways, airports, seaports, communications, water, power, even major buildings, can better optimize its resources, plan its preventive maintenance activities, and monitor security aspects while maximizing services to its citizens. Emergency response management to both natural as well as man-made challenges to the system can be focused and rapid. With advanced monitoring systems and built-in smart sensors, data can be collected and evaluated in real time, enhancing city management's decision-making. For example, resources can be committed prior to a water main break, salt spreading crews dispatched only when a specific bridge has icing conditions, and use of inspectors reduced by knowing condition of life of all structures. In the long term Smart Cities vision, systems and structures will monitor their own conditions and carry out self-repair, as needed. The physical environment, air, water, and surrounding green spaces will be monitored in non-obtrusive ways for optimal quality, thus creating an enhanced living and working environment that is clean, efficient, and secure and that offers these advantages within the framework of the most effective use of all resources [89].

There are a number of key elements needed to form a Smart City, and some of these are smart society, smart buildings, smart energy, smart lighting, smart mobility, smart water management etc. ICT forms the basic infrastructure; varying from sensors, actuators and electronic systems to software, Data, Internet and Cloud, Edge/fog and Mobile Edge computing. ICT is applied to improve these systems of systems building up a Smart City, making them autonomous and interoperable, secure and trusted. The interaction of the systems and the connectivity strongly depend on the communication gateway connecting the edge element data from sensors, actuators, and electronic systems to the Internet, managing- and control systems and decision programs. The communication gateway is a key enabler for the interconnection of systems in many applications such as Internet of Energy (IoE), Internet of Vehicles (IoV), Internet of Buildings (IoB) and Internet of Lighting (IoL). It is obvious with all the new systems and demand of interoperability that these communication gateways need more functionality, processing capacity, storage possibility, seamless connectivity, and more communication protocols embedded. At the same time the gateway must assure a higher level of security, interoperability and communication with devices across various verticals, such as energy, mobility and buildings.

An illustrative example is depicted in Figure 3.15 [76]. The Smart City is not only the integration and interconnection of intelligent applications, but also a people-centric and sustainable innovation model that is using communication and information technology and takes advantage of the open innovation ecology of the city and the new technologies such as IoT, cloud computing, smart data, and man-machine interaction.

A smart city is a developed urban area that creates sustainable economic development and high quality of life by excelling in multiple key areas: economy, mobility, environment, people, living, and government [105].

3.3.7.1 Large Scale Pilots and Ecosystem for Smart Cities

As main areas of application, smarter cities plays a relevant role, not only because the impact in re-using and re-purposing technology that is necessary (the number of deployed sensors) but also the increasing demand of new services (by citizens). IoT applications are currently based on multiple

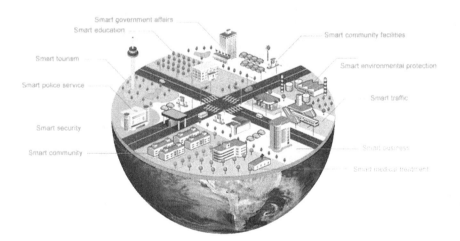

Figure 3.15 Smart City Concept [76].

architectures, technology standards and seamless software platforms, which have led to a highly fragmented IoT landscape. This fragmentation impacts directly the area of smart cities, which typically comprise several technological silos (i.e. IoT systems that have been developed and deployed independently for smart homes, smart industrial automation, smart transport, and smart buildings etc.).

A radical shift in the development, deployment and operation of IoT applications for smart cities, through introducing an abstract virtualized digital layer that operate across multiple IoT architectures, platforms (e.g. FI-WARE) and business contexts is required. Smart cities soon will face up the need for an integrated solution(s) (SmartCity-OS) that globally can monitor, visualise and control the uncountable integrated number of operations executed by diverse (and every day increasing) services platforms using the sensor technology deployed in the cities. Eventually this OS will be a blueprint across cities providing adaptive tools and generating the integration of other IoT systems and business opportunities.

Additional pointers to highlight are the quality of IoT Data and the numerous IoT Data source provisioning and the inherent need to generate

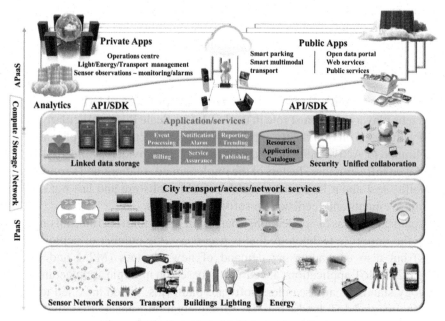

Figure 3.16   Smart City Multi-layered architecture.

semantic-driven business platforms, the reason to emphasize them is to address the emergent need for enabling business-driven IoT ecosystems and the generation of functionalities for Operating across multiple IoT architectures, platforms and business contexts, enable re(use) of Data for Smart Applications, and enable a more connected/integrated approach to smart city applications development.

There is a large way in the run towards integrated Internet of Things technological support and scientific progression towards interoperable connected objects, linked sensor data, pay-as-you-go IoT services and utility-driven privacy. Likewise the main areas to focus from a research perspective are, but not limited to, IoT architecture, systems and applications. The developments of IoT data modelling and schema representations, intra-domain and CPS extensions allow more robustness and extensible software platforms with embedded software and applications enabling Systems of Systems, peer-to-peer systems and applications.

An IoT large Scale Pilot is a fully designed, implemented and deployed ecosystem, such that all the players are inter-related: Technology Designers and Manufacturing, Software Designers and Developers, Research Institutions and Universities, Large Industries alike SME's, Alliance and standardization organisations, City Councils and Policy Makers and Citizen Organisations, share the same common objective which is sustainability. IoT Sustainability implies to have all the technology elements and services connected in the form of data interactions producing results.

3.3.7.2 Role of Institutions and Citizens in the Global IoT

The citizens play a protagonist role in the IoT Large Scale Pilot, particularly if the LSP is focused on Smart City applications. The role citizens can play are, but not limited to, Active elements in the system as data providers, Validation of the deployed infrastructure, Testers of the implemented solutions and services, Adaptability test about Robustness and Extensible software and last but not least Improvements and feedback on software solutions.

As main other areas of application, smart retail play a relevant role, not only because the impact in technology that is necessary (the number of deployed sensors) but also the increasing demand of new services (M2M and by citizens H2M). IoT applications are currently based on multiple architectures, technology standards and seamless software platforms, which have led to a highly fragmented IoT landscape. This fragmentation impacts directly the area of smart cities, which typically comprise several technological silos (i.e.

IoT systems that have been developed and deployed independently for smart homes, smart industrial automation, smart transport, and smart buildings etc.).

Excelling in these key areas can be done so through strong human capital, social capital, and/or ICT infrastructure. With the introduction of IoT a city will act more like a living organism, a city that can respond to citizen's needs.

In this context there are numerous important research challenges for smarty city IoT applications:

- Overcoming traditional silo based organization of the cities, with each utility responsible for their own closed world. Although not technological this is one of the main barriers.
- Creating algorithms and schemes to describe information created by sensors in different applications to enable useful exchange of information between different city services.
- Mechanisms for cost efficient deployment and even more important maintenance of such installations, including energy scavenging.
- Ensuring reliable readings from a plethora of sensors and efficient calibration of a large number of sensors deployed everywhere from lampposts to waste bins.
- Low energy protocols and algorithms.
- Algorithms for analysis and processing of data acquired in the city and making "sense" out of it.
- IoT large scale deployment and integration.

3.3.8 Smart Farming and Food Security

Food and fresh water are the most important natural resources in the world. Organic food produced without addition of certain chemical substances and according to strict rules, or food produced in certain geographical areas will be particularly valued. Similarly, fresh water from mountain springs is already highly valued. Using IoT in such scenarios to secure tracking of food or water from the production place to the consumer is one of the important topics.

The development of sensors, robots and sensor networks combined with procedures to link variables to appropriate farming management actions has open the opportunities for IoT applications in agriculture. The wired/wireless sensors, integrated into a IoT system can gather all the individual data needed for monitoring, control and treatment on (large scale) farms located in a particular region. This provides a mechanism of exchanging information in efficient ways enabling the execution of autonomously interventions in different agriculture sub-sectors (e.g. arable crops, livestock and horticulture).

IoT technology allows the monitoring and control of the plant and animal products during the whole life cycle from farm to fork. The challenge will be in the future to design architectures and implement algorithms that will support each object for optimal behaviour, according to its role in the Smart Farming system and in the food chain, lowering ecological footprint and economical costs and increasing food security.

The set of technologies used in smart farming is complex, to reflect the complexity of activities run by farmers, growers, and other sector stakeholders.

A recent report [85] on smart farming defines seven applications:

- Fleet management – tracking of farm vehicles
- Arable farming, large and small field farming
- Livestock monitoring
- Indoor farming – greenhouses and stables
- Fish farming
- Forestry
- Storage monitoring – water tanks, fuel tanks

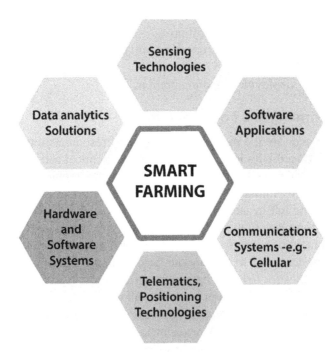

Figure 3.17 Different types of technologies involved in smart farming [85].

The report says that smart farming will allow farmers and growers to improve productivity and reduce waste, ranging from the quantity of fertiliser used to the number of journeys made by farm vehicles.

However, the complexity of smart farming is also reflected into the ecosystem of players. They can be classified in the following way:

- Technology providers – these include providers of wireless connectivity, sensors, M2M solutions, decision support systems at the back office, big data analytical systems, geo-mapping applications, smartphone apps.
- Providers of agricultural equipment and machinery (combines, tractors, robots), farm buildings, as well as providers of specialist products (e.g. seeds, feeds) and expertise in crop management and animal husbandry.
- Customers: farmers, farming associations and cooperatives.
- Influencers – those that set prices, influence the market into which farmers and growers sell their products.

The range of stakeholders in agriculture is broad, ranging from big business, finance, engineering, chemical companies, food retailers to industry associations and groupings through small suppliers of expertise in all the specialist areas of farming.

The end users of precision farming solutions include not only the growers but also farm managers, users of back office IT systems. Not to be forgotten is the role of the veterinary in understanding animal health. Also to be considered are farmers co-operatives, which can help smaller farmers with advice and funding.

The report concludes that the farming industry must embrace the IoT if it is to feed the 9.6 billion global population expected by 2050.

3.4 Future Internet Support for IoT

There are a number of challenges that the Future Internet community will need to address to adequately support the envisaged evolution of the Internet of Things. First we need to position these challenges within a 5–10 year timeline, and then introduce the technology enablers required to support the vision for future IoT-based applications and services. The following sections are reflecting three macro-challenges. One dedicated to the implications of having billions of connected "things" by 2020. The next one looking at what it takes to duly manage these connected devices in order to ensure dependable and robust services. The last one, more longer term oriented, will shed some light on what is required to usefully interpret the wealth of IoT harvested data and produce meaningful knowledge.

3.4.1 Macro-Challenges for Supporting IoT Evolution

As mentioned in the introduction, three macro-challenges have been identified as a suitable means to convey the main implications for the Future Internet, derived from the evolution of IoT.

The first one relates to the already ongoing trend of having more and more devices and more generically "objects/things" connected to the Internet. Forecasts vary in numbers according to who made the predictions and what those predictions entailed. There is however no disagreement on the fact that there will be billions of connected objects by 2020. The sheer scale of connected devices and the type of traffic these generate (compared to human's devices) will have substantial implications on the Internet as we know it today.

Managing objects, and ensuring that they can be seamlessly integrated in different application domains and ensuring that the data they produce can be reliably accessed to sustain dependable services is part of the second macro-challenge identified. There are currently many IoT services being used, though mostly perceived as "best effort" due to the nature of the resources involved (end devices that get out of coverage, out of battery, jammed through interference, need for human intervention for configuring, replacing, maintaining them etc.).

This second macro-challenge is concerned with supporting the "tactile Internet" which by many is already being hailed as the natural evolution of IoT. Sensing substrates and communication infrastructures are getting more tightly geared towards supporting more agile and reactive applications.

Getting billions of objects duly connected and managing these to create a reliable monitoring/actuating substrate only partially caters for the challenges ahead. These challenges cannot be complete without considering how to handle the huge amount of data produced and how to transform it into useful and actionable knowledge. This is indeed the most difficult of the macro-challenges ahead given it is related to intelligent reasoning over the data IoT will produce. The difficulty of this challenge lies in the lack of general purpose machine-learning based solutions that can be re-used to address the wide variety of situations in which similar IoT services and applications could be applied.

The figure below illustrates a visual map of these macro-challenges, together with the associated sub-challenges as illustrated in the next section.

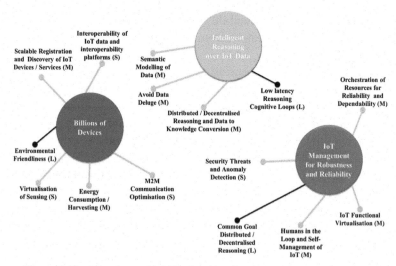

Figure 3.18 Visual map for IoT-related implications of the Future Internet.

The letters S, M and L are there to indicate the Short, Medium or Long term nature of the sub-challenge.

3.4.1.1 Billions of Devices

Scalable registration and discovery of IoT devices/services – As more and more devices get connected, the challenge becomes finding them in a context-aware way. Semantic enrichment of objects description is poised to play a role in facilitating the automated discovery of suitable objects for the purposes of the various applications.

Interoperability of IoT data and interoperability platforms – Besides the common aspects of underlying technologies that have enabled short-range connectivity and the miniaturization of devices that have paved the way towards the success of IoT, current applications have been evolving as a collection of vertical silos often deployed with different standards. To fully unlock the potential of having billions of connected objects, cross-use of data across application domains will be needed. Solutions that foster interoperability and reduce barriers between application silos will therefore have a strong role to play.

Virtualisation of sensing – Following on the need to foster interoperability, virtualization of objects will also be needed to separate real and resource-constrained objects from their virtual counterparts in order to minimising energy consumption, facilitate interaction with applications as well as address the challenges of scalability and those of empowering single objects with flexible added resources from the "wired and resource intensive world".

M2M communication optimisation – Connected objects have communication requirements that can be substantially different from devices like computers or smart-phones. Short-lived communications in huge numbers and energy consideration will require redesign of communication protocols, especially for the wireless part to minimise the overheads associated with exchanging data between objects and their corresponding clients/gateways.

Energy consumption/harvesting – To ensure long duration and usefulness of connected objects, given also limitations of battery evolution compared to processing power and spectrum efficiency, it will be essential to design hardware and systems that can operate for long time without need for battery replacement/recharging. Integration of energy harvesting techniques also falls in this category.

Environmental friendliness – Billions of devices lasting up to 5–10 years, but very often replaced much earlier, means a lot of waste produced after these devices are no longer operational. Choice of fully recyclable materials fostering sustainability of IoT will be more and more important. Especially after the many IoT deployments will produce sustained need of hardware and services and differentiation between vendors will start including these environmental friendliness factors.

3.4.1.2 IoT Management for Robustness and Reliability

Security threats and anomaly detection – This is a cross-cutting issue as it relates not only to the security of radio communications, but also to the security of IoT-generated data to ensure good levels of trust and privacy. On this front not only solutions that address these issues are needed but also solutions that at a management level can detect attacks and contain them.

Orchestration of resources for reliability and dependability – This challenge relates to the ability of assessing dependencies between sensing, networking and computing resources and how these components contribute to the QoE and reliability of the end-to-end application being supported. Issue of dependability becomes important if one has to leverage on the advantages of the IoT also within mission critical systems and/or simply more dependable services.

IoT function virtualisation – The IoT functionality is currently solely supported by ad-hoc hardware (i.e. communication of sensed-data, domain/sensor specific gateways etc.). IoT function virtualisation will be opening up new opportunities where hardware ownership will not be necessarily a requirement for producing IoT services.

Common goal distributed/decentralised reasoning – As IoT functionality gets virtualised and distributed besides orchestrating the use of resources there will also be need to coordinate decision-making and achieve conflict resolution for the actuators that are involved in achieving a common goal.

3.4.1.3 Intelligent Reasoning over IoT Data

Semantic modeling of data – As more and more data gets collected through IoT devices, to ensure a more automated selection of the appropriate end devices to be associated with IoT services and applications, IoT data will have to be modeled according to given structures and properly annotated. Semantics help in this respect; so this challenge is part of the broader data interoperability problem though it encompasses besides "finding" the right data also the ability of fostering automated translation between data structures in different ontology domains.

Avoid data deluge – This challenge is about the ability of processing data close to the place where it is generated or on its way to the requesting application. This will help avoid unnecessary use of network resources, as well as reduce the amount of data that have to be processed for analytics purposes. It includes challenges like data aggregation, stream processing, CEP etc.

Distributed/decentralised reasoning and data to knowledge conversion – While the previous challenge is about why we should avoid data deluge, this challenge is about how this can be achieved. IoT is becoming the underlying monitoring fabric of future smart-x applications. Trends suggests there will soon be more devices than we can dedicate attention to, thus getting data across to applications will have to be better managed on the end-to-end delivery path. This requires introduction of new ways for distributed data interpretation which accounts for the locality of data, the need to compress it to meet application requirements (i.e. latency, quality etc.) and network capacity.

Low latency reasoning cognitive loops – This challenge relates to the IoT evolving towards becoming able to support very low-latency reasoning loops. This involves the ability to instantiate data processing instances dynamically and close to data sources, besides addressing redesign of communication protocols for speed.

Humans in the loop and self-management of IoT – The rapidly increasing number of connected objects will not be met by a similarly progress in humans ability to set them up, configure them, manage them etc. This element of the roadmap relates to the need of solutions that will ensure that devices can be fully operational with simple and little involvement of the users, if need be.

3.4.2 Roadmap and Technology for Addressing These Challenges

IoT is currently positioned at the top of the Gartner hype cycle (the so called "peak of expectations"). The challenge for all the businesses that plan to draw on the wide uptake of this technology, is to ensure that the "through of disillusionment" is somewhat reduced and that the market remains sustained. To achieve this objective one must certainly focus on adoption and user-friendliness. From a technology viewpoint, effort should go to ensuring that the right enablers that support this vision are developed.

3.4.2.1 From Challenges to Technology Solutions

In previous section we split the future IoT challenges under three main categories, one related to dealing with billions of connected things, one related to having to manage these devices and a last one associated with making the most of data these things will produce in other words, how to create useful knowledge.

What is clear is that full mesh connectivity between billions of devices and associate applications will not be achievable, which brings us to the statement that IoT will need increased flexibility in the "communication infrastructure substrate".

This translates into ensuring adequate evolution in the following technology domains: flexible networks for prioritized and M2M-specific communications, edge cloud computing and distributed big-data analytics. Besides these "infrastructure oriented" technologies, also progress in more "IoT specific" domains will be needed: this relates to progress on the "hardware-related" energy harvesting side to ensure more reliable and durable IoT services. Whereas on the "software side" semantic technologies as well as ensuring security and privacy protection solutions need to be reliable and usable to foster wide acceptance.

The remainder of this section sheds more light into the technologies (light boxes in the figures below) that support the highlighted challenges (dark boxes in the figures). As before, we keep the similar structure around the three earlier presented challenges.

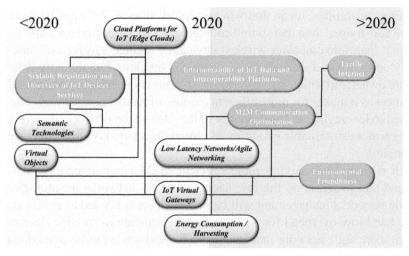

Figure 3.19 Billions of connected devices.

Billions of connected devices, what technology enablers? – Addressing the challenge of billions of connected devices will certainly require some scale-proof technologies for enabling their automated registration, search and discovery, maintenance and management.

This has a lot to do with progress on the semantic technologies front (and subsequent semantic annotation of objects). Through the use of semantics one can design how to automatically relate all devices that e.g. share a similar location, or that can produce a certain type of data, or that are owned by the same person. Moreover progress on the semantic technologies front is also needed to address IoT application silos interoperability problems. In particular this will enable the system's understanding of "what" needs to be done to achieve interoperability between data in separate domains. As far as the "how" is concerned, once it is clear what conversion needs to be applied to the sensed data to make it available for e.g. across application domains. Here comes the role of edge clouds where appropriate algorithms can be instantiated and run to address interoperability issues.

Looking at more "hardware" related issues, progress in the energy harvesting field has many implications on the achievement of the billions devices challenge. It certainly contributes to environmental friendliness as it relies upon renewable energy for the installation of sensing devices at zero energy impact i.e. without connection to power sources. Similarly, the relatively slow advances in battery technologies compared to evolution of computing capabilities, mean that wireless IoT devices will always be more resource constrained than their wired counterparts. Virtualisation (of sensing) techniques therefore empower wireless devices by adding "always-on" functionality on the "wired side" of the network and breaking functionality from hardware ownership which also contributes to achieving better environmental friendliness as it makes for more efficient (re-)use of hardware resources. This is aligned to leveraging on functionality to the edge of the network, therefore enabling a more sustainable evolution of virtual objects/IoT Virtual Gateway functionality.

With regards to achieving more efficient M2M communications, it is envisaged that progress on the low-latency wireless networks technologies and agile networks management will be needed. This is needed to ensure on the one hand low-overhead for short-lived communications to edge devices (through more agile network management schemes) while on the other hand achieving shorter cognitive loops for sensing-processing-actuating close to the edge, a "must-have" requirement for tactile internet future scenarios.

Management of IoT devices for robustness and reliability – The importance of this challenge stems from IoT becoming more mature and established enabling contextually also support for critical services, or more robust and dependable ones in general.

From a technology viewpoint, there is need for more flexible "infrastructure oriented" technologies to mature. Edge clouds and software networks

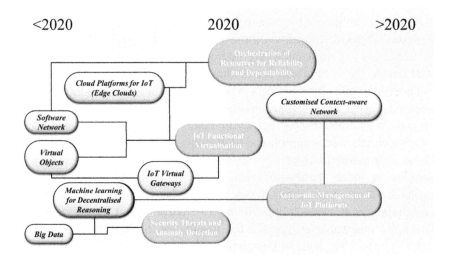

Figure 3.20   Management of IoT for robustness and reliability.

are there to support and complement the constrained nature of devices and have therefore implications on the sub-challenges of virtualizing IoT functionality and orchestrating the use of these "infrastructure technologies" for a more robust IoT. Virtual Objects and Virtual Gateways are also specific IoT technologies, building bricks of virtual IoT functions which can be more robust and resilient to connected objects hardware failures/limited coverage.

With an increase of the number of devices beyond what humans can successfully manage comes the need to rely on cognitive technologies for autonomic management of IoT platforms and for security threats and anomaly detection. Specifically, this is supported by progress in the big-data analytics and leverages on machine learning and decentralized reasoning technologies.

Intelligent reasoning over IoT data – While previous challenges were related to IoT hardware and more infrastructure oriented, this one is about how to best leverage on IoT harvested data, notably to produce the usable and useful knowledge for compelling IoT-based services and applications in many different domains.

Semantic annotation of data is a must to be able to automatically draw "relevance boundaries" amongst available data. Hence, progress on semantic technologies underpins the development of data models that foster and support well-targeted data to knowledge conversions which is key in ensuring wide adoptions (i.e. cognitive systems that take the right decisions through predictive models).

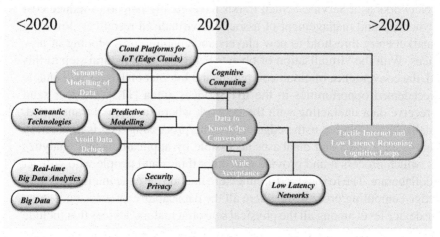

Figure 3.21 Intelligent reasoning over IoT data.

Besides semantic technologies, techniques of cognitive computing are also required. Here we refer to the more and more reliable services that large computing machines such as e.g. IBM Watson will enable. The algorithms for data to knowledge manipulation and for predictive modelling contribute to better quality decisions and wide acceptance. This is also where big-data steps-in, as well as security and privacy by design which are also key to ensure wide acceptance.

On the "data to knowledge conversion" path we illustrated the importance of real-time big-data analytics applied to reduce in size the produced IoT data, and thus lowering IoT impact on communication networks. This is achieved through pre-processing done close to the edge, avoiding data deluge.

Edge clouds and low latency networks, together with well-targeted data to knowledge conversion are the key technologies for achieving fast reasoning loops, which underpin future Tactile Internet scenarios.

3.5 Internet of Things and Related Future Internet Technologies

3.5.1 Cloud and Edge/Fog Computing

Cloud computing has been established as one of the major building blocks of the Future Internet. New technology enablers have progressively fostered virtualisation at different levels and have allowed the various paradigms known as "Applications as a Service", "Platforms as a Service" and "Infrastructure and Networks as a Service". Such trends have greatly helped to reduce cost of ownership and management of associated virtualised resources, lowering the market entry threshold to new players and enabling provisioning of new services. With the virtualisation of objects being the next natural step in this trend, the convergence of cloud computing and Internet of Things will enable unprecedented opportunities in the IoT services arena [112]. Devices send and receive data interacting with the network where the data is transmitted, normalized, and filtered using edge computing/processing then is transferred in data storage units and databases accessible by applications and analytics tools, which process it and provide it to other things and people who will act and collaborate. The IoT layered architecture include the edge intelligence into the edge computing/processing where all the data capture, processing is done at the device level among all the physical sensor/actuators/devices that include controllers based on microprocessors/microcontrollers to compute/process and wireless modules to communicate. The intelligence at the edge supports devices to use their data sharing and decision-making capabilities to interact and cooperate in order to process the data at the edge, filter it and select/prioritize what is important. This intelligent processing at the edge select the "smart data" that is transferred to the central data stores for further processing in the cloud. This allows including the Edge Cloud for processing data and addressing the challenges of response-time, reliability and security. For real time fast processes, the sensor/actuator edge devices could generate data much faster than the cloud-based apps can process it.

The use of intelligent edge devices require to reduce the amount of data sent to the cloud through quality filtering and aggregation and the integration of more functions into intelligent devices and gateways closer to the edge reduces latency. By moving the intelligence to the edge, the local devices can generate value when there are challenges related to transferring data to the cloud. This will allow as well for protocol consolidation by controlling the various ways devices can communicate with each other.

As part of this convergence, IoT applications (such as sensor-based services) will be delivered on-demand through a cloud environment [113]. This extends beyond the need to virtualize sensor data stores in a scalable fashion. It

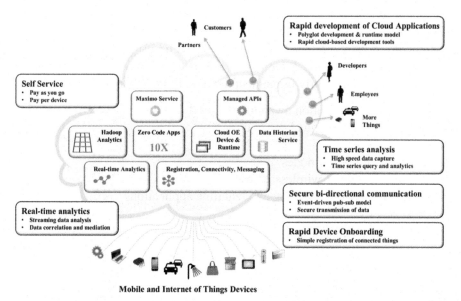

Figure 3.22   Internet of Things Cloud (Source: IBM).

asks for virtualization of Internet-connected objects and their ability to become orchestrated into on-demand services (such as Sensing-as-a-Service).

With the growth of IoT, we're shifting toward a cyber-physical paradigm, where we closely integrate computing and communication with the connected things, including the ability to control their operations. In such systems, many security vulnerabilities and threats come from the interactions between the cyber and physical domains. An approach to holistically integrate security vulnerability analysis and protections in both domains will become increasingly necessary. There is growing demand to secure the rapidly increasing population of connected, and often mobile, things. In contrast to today's networks, where assets under protection are typically inside firewalls and protected with access control devices, many things in the IoT arena will operate in unprotected or highly vulnerable environments (i.e. vehicles, sensors, and medical devices used in homes and embedded on patients). Protecting such things poses additional challenges beyond enterprise networks [60].

Many Internet of Things applications require mobility support and geo-distribution in addition to location awareness and low latency, while the data need to be processed in "real-time" in micro clouds or fog. Micro cloud or Fog computing enables new applications and services applies a different data management and analytics and extends the Cloud Computing paradigm to the edge of the network. Similar to Cloud, Micro Cloud/Edge Cloud/Fog provides data, compute, storage, and application services to end-users.

The Micro Cloud or the Edge Cloud/Fog needs to have the following features in order to efficiently implement the required IoT applications:

- Low latency and location awareness
- Wide-spread geographical distribution
- Mobility
- Very large number of nodes
- Predominant role of wireless access
- Strong presence of streaming and real time applications
- Heterogeneity

The worlds of IT and telecommunications networking are converging bringing with them new possibilities and capabilities that can be deployed into the network. A key transformation has been the ability to run IT based servers at network edge, applying the concepts of cloud computing. This is defined as Mobile Edge computing [69]. Mobile edge computing can be seen as a cloud server running at the edge of a mobile network and performing specific tasks that could not be achieved with traditional network infrastructure. IoT/M2M gateway and control functions are typical examples, but there are many others. Mobile edge computing is characterized by [69]:

- On-Premises: The Edge is local, meaning that it can run isolated from the rest of the network, while having access to local resources. This becomes particularly important for M2M scenarios, for example when dealing with security or safety systems that need high levels of resilience.
- Proximity: Being close to the source of information, Edge Computing is particularly useful to capture key information for analytics and big data. Edge computing may also have direct access to the devices, which can easily be leveraged by business specific applications.
- Lower latency: As Edge services run close to end devices it considerably reduces latency. This can be utilized to react faster, to improve user experience, or to minimize congestion in other parts of the network.
- Location awareness: When a Network Edge is part of a wireless network, whether it is Wi-Fi or Cellular, a local service can leverage low-level signalling information to determine the location of each connected device. This gives birth to an entire family of business-oriented use cases, including Location Based Services, Analytics, and many more.
- Network context information: Real-time network data (such as radio conditions, network statistics, etc.) can be used by applications and services to offer context-related services that can differentiate the mobile broadband experience and be monetized. New applications can be developed (which will benefit from this real-time network data) to connect mobile subscribers with local points-of-interest, businesses and events.

Mobile Edge computing transforms base stations into intelligent service hubs that are capable of delivering highly personalized services directly from the very edge of the network while providing the best possible performance in mobile networks. Proximity, context, agility and speed can be translated into unique value and revenue generation, and can be exploited by operators and application service providers to create a new value chain [69].

For the future IoT applications it is expected that more of the network intelligence to reside closer to the source. This will push for the rise of Edge Cloud/Fog, Mobile Edge computing architectures, as most data will be too noisy or latency-sensitive or expensive to be transfer to the cloud.

3.5.2 Federated IoT Data Cloud and Orchestration of Large Scale Services

The rapid evolution of Sensor Technologies, the Semantic Web consolidation and the extensive deployment of Cloud Computing Systems provide a unique opportunity to unify the real and the virtual worlds (Internet of Things). The Internet of Things enables the building of very large infrastructures that for the first time facilitate the information-driven real-time integration of the physical world and computers (Cyber Physical Systems) on a global scale (connecting sensor with systems and systems with the web). At the same time IoT can be considered a flexible middleware technology that abstracts from heterogeneous sensor network technologies to higher-level functionalities to enable interconnected sensor networks and processing of sensor data (Sensor Internet). It is a cornerstone for enabling Semantic Reality.

Cloud computing comprises the computing capacity necessary to run background operations to facilitate the complex IoT data analytics. The vision towards a Global Internet of Things requires not only emergent technologies but heterogeneous IoT system of systems coordinated via Federated services platforms. The deployment of IoT Data Cloud management systems and the orchestration of Large Scale Services are also important to enable a "global" view of the services and IoT infrastructures. In a global Internet of Things, control of sensors or infrastructure became a secondary role as per the Internet of Things services creation mechanisms are focused on providing capabilities and functionality and repurposing services and data rather than configurations and infrastructure adaptation/changes.

The sheer size of global Internet of Things systems pose novel and unique challenges, as it can only be engineered and deployed if a large degree of self-organization and automation capabilities are offered (large-scale deployments). Global internet of Things are built into the system and its constituents, enabling simple deployment (plug-and-play), dynamic (re-)configuration, re-purposing of technology and flexible component and information integration alike tailored information delivery based on user context and needs in a service-oriented way. This requires semantic descriptions of the user needs and contexts, and of the system's constituents, the data streams they produce, their functionalities and their requirements to enable a machine-understandable information space of real-world entities and their dynamic communication processes on a scale that is beyond the current size of the Internet.

IoT is expanding rapidly and is changing the perception of our daily life, not only from a technological perspective but also our personal activities, professional career and also in the way we establish social interactions. IoT is already considered as crucial in the process for designing the Future Internet. It is expected IoT will revolution our perception of the world enabling more smartness to the different external aspects of the human being (cities, industries, agriculture, clothing, fashion, etc.)

Currently Internet of Things not only has planned the model for "global" distributed infrastructures worldwide interconnected but envisioned the creation of distributed applications that rely in non-proprietary technologies (e.g. Web of Things, Internet of Everything, and the Physical Web). The adoption of IoT technology and its immersion in the society is generating high demands for high volumes of data and the capacity for storing, processing and analysing it in real time.

Based on the evolution of sensor technologies and the semantic technologies to unify the real and the virtual worlds this global vision is becoming a reality. It is yet a need for investigating the convergence of systems and technology platforms (e.g. software systems, the semantic web technologies and the Internet). The main objective is for developing flexible IoT middleware solutions/technology which abstracts data from heterogeneous sensor networks and bring this to a higher application-level(s) for enabling extended systems' functionalities and also enable interconnected sensor networks and sensor web data interoperability.

Extensive attention is necessary to focus on the deployment, maintenance and monitoring work on large-scale deployments, big sensor data collection and annotation and investigate data transformation and processing by means of advance stream processing techniques. To this end it is necessary to work on the design principles for device and infrastructure-related architectures, technologies and protocol frameworks for Internet connected heterogeneous devices.

3.5.2.1 IoT Data Analytics

The need for efficient Methods and Algorithms for Big Data, Collection and Transformation following self-Organization and self-Management paradigms still remains as one of the main objectives in the evolution towards Global Internet of Things. Cloud Computing Infrastructures and Management Platforms have evolved but Privacy and Security-Enabled Middleware Platforms are expected research activities in relation to topics that are not limited to Cloud Infrastructures for Data Analytics, Security, Privacy and Trust, Recommender

Systems and Clustering Mechanisms, Federation and Orchestration, Service Configuration and Control, Ontology Engineering and Applied Semantics alike Modelling and Reasoning Techniques.

A major effort on investigating the convergence of software systems, the semantic web and the Internet, heavily focused on the evolution of sensor technologies and the semantic technologies to unify the real and the virtual worlds. Development of flexible IoT middleware technology which abstracts data from heterogeneous sensor networks and bring this to a higher application-level(s) for enabling extended systems' functionalities and also enable interconnected sensor networks and sensor web data interoperability by using the Internet. Extensive work on large-scale deployments, big sensor data collection and annotation is due to come and investigate data transformation and processing by means of advance stream processing techniques query languages and reasoning techniques for the amount of generated data in the city. Design efforts for defining principles for device and infrastructure-related architectures, technologies and protocol frameworks for Internet connected heterogeneous devices.

Based on the evolution of sensor technologies and the semantic technologies it is possible to unify the real and the virtual worlds. There is yet the need for investigating the convergence of systems and technology platforms (e.g. software systems, the semantic web technologies and the Internet). The main objective is for developing flexible IoT middleware solutions/technology which abstracts data from heterogeneous sensor networks and bring this to a higher application-level(s) for enabling extended systems' functionalities and also enable interconnected sensor networks and sensor web data interoperability by using the Internet. Extensively is necessary to focus on the deployment, maintenance and monitoring work on large-scale deployments, big sensor data collection and annotation and investigate data transformation and processing by means of advance stream processing techniques. To this end it is necessary to work on the design principles for device and infrastructure-related architectures, technologies and protocol frameworks for Internet connected heterogeneous devices.

3.5.3 IoT Interoperability and Semantic Technologies

The previous IERC SRIAs have identified the importance of interoperability semantic technologies towards discovering devices, as well as towards achieving semantic interoperability.

Interoperability is defined as the ability of two or more systems or components to exchange data and use information this provides many challenges on how to get the information, to exchange data, and to understand and process the information. There are four basic IoT interoperability layers to be thoroughly tested and validated: technical, syntactical, semantic, and organizational.

- Technical Interoperability is usually associated with hardware/software components, systems and platforms that enable machine-to-machine communication to take place. This kind of interoperability is often centred on (communication) protocols and the infrastructure needed for those protocols to operate.
- Syntactical Interoperability is usually associated with data formats. Certainly, the messages transferred by communication protocols need to have well-defined syntax and encoding, even if it is only in the form of bit-tables. However, many protocols carry data or content, and this can be represented using high-level transfer syntaxes such as HTML, XML or ASN.1.
- Semantic Interoperability is usually associated with the meaning of content and concerns the human rather than machine interpretation of the content. Thus, interoperability on this level means that there is a common understanding between people of the meaning of the content (information) being exchanged.
- Organizational Interoperability is the ability of organizations to effectively communicate and transfer (meaningful) data (information) even though they may be using a variety of different information systems over widely different infrastructures, possibly across different geographic regions and cultures.

Organizational interoperability depends on the former three. Following the definitions and the trends on ICT sector about sensors and sensor data we can add two other dimensions: Static and dynamic interoperability.

- Dynamic interoperability: Two products cannot interoperate if they do not implement the same set of options ("services"). Therefore when specifications are including a broad range of options, this aspect could lead to serious interoperability problem. Solutions to overcome these aspects consist of definition clearly in a clear document the full list options with all conditions (e.g. defined as PICS in ISO 9646 [77]) as well as to define set of profiles. In the latter case, defining profile would help to truly check interoperability between two products in the same family or from different family if the feature checked belongs to the two groups. We could consider this aspect as

- Static interoperability using approach of the well-known OSI overall test methodology ISO 9646 [77], where there is definition of static conformance review. Conformance testing consists of checking whether an Implementation Under Test (ITU) satisfies all static and dynamic conformance requirements. For the static conformance requirements this means a reviewing process of the options (PICS) delivered with the IUT. This is referred to as the static conformance review. This aspect could appear easy but that represent serious challenge in the IoT field due the broad range of applications.

The solutions that use non-interoperable solutions lead to increase of complexity in communicating and interpreting their data and services. One interesting research is to accept differences and potential non-interoperability for instance between two different protocols but to adapt on the fly. We see also such features in intelligent gateways and middleware. This can be called dynamic interoperability and should be a continuous important research area in particular with the growing complexity and heterogeneity of IoT environments.

The challenges for IoT interoperability are many and there is a need for an interoperability framework to address them in a consistent manner under the IoT architectural model. These challenges require addressing a number of research topics as presented in Table 3.1.

Table 3.1   IoT Interoperability research topics

Challenges	Research Topics
Discovery of objects and Clustering	• Algorithms for data selection and classification • Efficient clustering mechanisms • IoT service management systems
Privacy and Security at Technical and Semantic level	• Access control algorithms and tools • Rules-based systems • IoT systems federation
Quality of Data	• Data filtering and data selection • Data mining • Control and assurance
Reasoning and Analysis	• Taxonomy, modelling, • Probabilistic modelling • Inference, Abstraction and Abduction
Data Management	• Data fusion • Mash-ups processing • Stream processing

There are arguments against using semantics in constrained environments since ontologies and semantic data can add too much overhead in the case of devices with limited resources. However, ontologies are a way to share and agree on a common vocabulary and knowledge; at the same time there are machine-interpretable and represented in interoperable and re-usable forms. There is no need to add semantic metadata in the source, since this could be added to the data at a later stage (e.g. in a gateway that have mere functionalities). The legacy applications can ignore these ontologies or can be extended to work with it.

In IoT applications semantic technologies will have an important role in enabling sharing and re-use of virtual objects as a service through the cloud. The semantic enrichment of virtual object descriptions will realise for IoT what semantic annotation of web pages has enabled in the Semantic Web. Associated semantic-based reasoning will assist IoT users to more independently find the relevant proven virtual objects to improve the performance or the effectiveness of the IoT applications they intend to use.

3.6 Networks and Communication

Present communication technologies span the globe in wireless and wired networks and support global communication by globally-accepted communication standards. The Internet of Things Strategic Research and Innovation Agenda (SRIA) intends to lay the foundations for the Internet of Things to be developed by research through to the end of this decade and for subsequent innovations to be realised even after this research period. Within this timeframe the number of connected devices, their features, their distribution and implied communication requirements will develop; as will the communication infrastructure and the networks being used. Everything will change significantly. Internet of Things devices will be contributing to and strongly driving this development.

Changes will first be embedded in given communication standards and networks and subsequently in the communication and network structures defined by these standards.

3.6.1 Networking Technology

Mobile traffic today is driven by predictable activities such as making calls, receiving email, surfing the web, and watching videos. Over the next 5 to 10 years, billions of IoT devices with less predictable traffic patterns will join

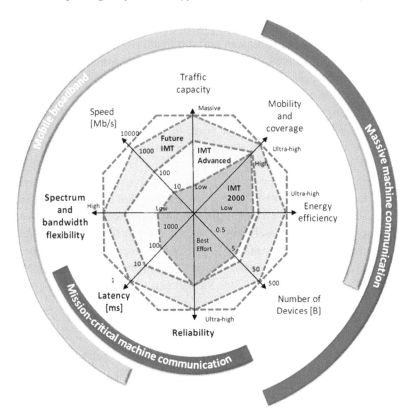

Figure 3.23   Global perspective – 5G capabilities [157].

the network, including vehicles, machine-to-machine (M2M) modules, video surveillance that requires 24-7 bandwidth, or different types of sensors that send out tiny bits of data each day. The rise of cloud computing requires new network strategies for fifth evolution of mobile the 5G, which represents clearly a convergence of network access technologies. The architecture of such network has to integrate the needs for IoT applications and to offer seamless integration. To make the IoT and M2M communication possible there is a need for fast, high-capacity networks.

The capabilities depicted in Figure 3.23 are the following [157]:

- *Traffic capacity* relates to the capability to manage a certain amount of offered traffic per area unit.
- *Mobility/coverage* refers to the capability to provide connectivity in any situation; on the move and when standing still, regardless of user location.

- *Network and device energy efficiency* relates to the energy consumption in both wireless devices and network infrastructure.
- *Massive number of devices* relates to the capability to handle a large number of connected devices per area unit, while preventing that the related control signalling overhead limits the user experience.
- *Reliability* relates to the capability to provide a given service level with very high probability. If reliability is high enough, mission-critical and safety-of-life applications can be supported.
- *Latency* refers to the time the system needs to transport data through its own domain of responsibility.
- *Spectrum and bandwidth flexibility* refers to the flexibility of the system design to handle different spectrum scenarios, and in particular to the capability to handle higher frequencies and wider bandwidths than today.
- *Achievable end user data rate* refers to the maximum data rate a user typically experiences (i.e. the "perceived speed" of the data connection).

The capabilities relate to the use cases for future international mobile telecommunications, as shown through the arches at the edge of the figure.

- Mobile Broadband is the human centric use case for non-limited access to services and data anytime and anywhere.
- Mission-critical machine communication is a use case where communication between machines is required to have an exactly defined behaviour in terms of key KPIs such as guaranteed throughput, latency, etc. Examples are wireless control of industrial manufacturing or production processes, traffic safety applications, etc.
- Massive Machine Communication is a use case mainly characterized by a very large number of connected devices which typically transmit relatively low volume of non-delay-sensitive data. Devices are required to be simple and cheap, and have a very long battery life.

5G networks will deliver 1,000 to 5,000 times more capacity than 3G and 4G networks today and will be made up of cells that support peak rates of between 10 and 100 Gbps. They need to be ultra-low latency, meaning it will take data 1–10 milliseconds to get from one designated point to another, compared to 40–60 milliseconds today. Another goal is to separate communications infrastructure and allow mobile users to move seamlessly between 5G, 4G, and WiFi, which will be fully integrated with the cellular network. Networks will also increasingly become programmable, allowing operators to make changes to the network virtually, without touching the physical infrastructure. The capabilities of Future and previous future

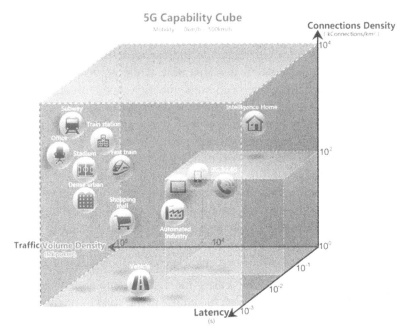

Figure 3.24 5G Capability cube (Source: Ericsson).

international mobile telecommunications systems are depicted in Figure 3.24. Future international mobile telecommunications will encompass all the capabilities of the previous systems. The performance requirements in some scenarios will increase significantly due to new services arising and spreading.

The evolution and pervasiveness of present communication technologies has the potential to grow to unprecedented levels in the near future by including the world of things into the developing Internet of Things. Network users will be humans, machines, things and groups of them.

3.6.2 Communication Technology

The growth in mobile device market is pushing the deployment of Internet of Things applications where these mobile devices (smart phones, tablets, etc. are seen as gateways for wireless sensors and actuators.

Communications technologies for the Future Internet and the Internet of Things will have to avoid such bottlenecks by construction not only for a given status of development, but for the whole path to fully developed and still growing nets.

Many types of Internet of Things devices will be connected to the energy grid all the time; on the other hand a significant subset of Internet of Things devices will have to rely on their own limited energy resources or energy harvesting throughout their lifetime.

The inherent trend to higher complexity of solutions on all levels will be seriously questioned – at least with regard to minimum energy Internet of Things devices and services.

Their communication with the access edges of the Internet of Things network shall be optimized cross domain with their implementation space and it shall be compatible with the correctness of the construction approach.

The next years' M2M associated with the Internet of Things could be SIM-*less*, meaning "wireless, long-range, low-power, low data-rate and without SIM-card". A deep revolution in the landscape of M2M wireless radio communication technologies is taking off. Until now, in the field of M2M, only GPRS, SMS, 3G technologies based on the SIM card principle allowed to pass information over long distances between an object and a remote information system. Once the SIM card is integrated in the sensor, the object becomes communicating. It can be fixed (drinks vending machine, tank, thermostat, energy box, smoke detector, parking meter) or mobile (wagons, containers, heavy vehicles, bicycles). A sensor then records data locally and transmits it automatically via the integrated GSM modem to the remote information system.

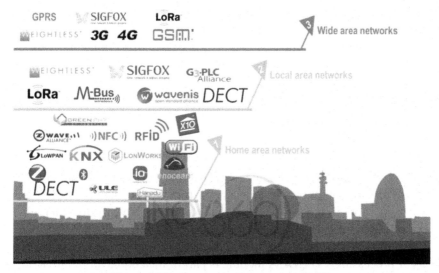

Figure 3.25 Communication standards [158].

Innovative wireless communication technologies create new perspectives for IoT applications. Companies such as Sigfox (French IoT operator using its own technology), Weightless (promoted mainly in UK), LoRa (french Cycleo technology acquired by the founder Semtech) are introducing these wireless communication technologies. This new technologies push for the emergence of new attractive and alternative business models for certain IoT applications with subscriptions ranging from 1 € to 20 € per year without cost of data. While these solutions can only transmit a small amount of data per message (dozens or even hundreds of kilobytes per message), they however cover well over 80% of M2M and IoT's needs. Some applications have already rejected the SIM-card GSM approach to precisely focus on SIM-*less*. This is the case for smart meters. Except for market's structural cause, over the next eight years, nearly 200 million residential meters (Water, Gas, Electricity) across the members of the Euro zone countries will be connected and half will be equipped with SIM-*less* technology (and the other half in PLC) [158].

3.7 Data Management

Data management is a crucial aspect in the Internet of Things. When considering a world of objects interconnected and constantly exchanging all types of information, the volume of the generated data and the processes involved in the handling of those data become critical.

In this context there are many technologies and factors involved in the "data management" within the IoT context.

Some of the most relevant concepts which enable us to understand the challenges and opportunities of data management are:

- Data Collection and Analysis
- Big data
- Semantic Sensor Networking
- Virtual Sensors
- Complex Event Processing

Data Collection and Analysis modules or capabilities are the essential components of any IoT platform or system, and they are constantly evolving in order to support more features and provide more capacity to external components (either higher layer applications leveraging on the data stored by the DCA module or other external systems exchanging information for analysis or processing). The DCA module is part of the core layer of any IoT platform.

An example of data management framework for IoT is presented in [73] that incorporates a layered, data-centric, and federated paradigm to join the independent IoT subsystems in an adaptable, flexible, and seamless data network. In this framework, the "Things" layer is composed of all entities and subsystems that can generate data. Raw data, or simple aggregates, are then transported via a communications layer to data repositories. These data repositories are either owned by organizations or public, and they can be located at specialized servers or on the cloud. Organizations or individual users have access to these repositories via query and federation layers that process queries and analysis tasks, decide which repositories hold the needed data, and negotiate participation to acquire the data. In addition, real-time or context-aware queries are handled through the federation layer via a sources layer that seamlessly handles the discovery and engagement of data sources. The whole framework allows a two-way publishing and querying of data. This allows the system to respond to the immediate data and processing requests of the end users and provides archival capabilities for later long-term analysis and exploration of value-added trends.

In the context of IoT, data management systems must summarize data online while providing storage, logging, and auditing facilities for offline analysis. This expands the concept of data management from offline storage, query processing, and transaction management operations into online-offline communication/storage dual operations. The lifecycle of data within an IoT system is illustrated in Figure 3.26, proceeds from data production to aggregation, transfer, optional filtering and pre-processing, and finally to storage and archiving. Querying and analysis are the end points that initiate (request) and consume data production, but data production can be set to be "pushed" to the IoT consuming services. Production, collection, aggregation, filtering, and some basic querying and preliminary processing functionalities are considered online, communication-intensive operations. Intensive pre-processing, long-term storage and archival and in-depth processing/analysis are considered offline storage-intensive operations [73].

The proposed IoT data management framework consists of six stacked layers, two of which include sub-layers and complementary or twin layers. The framework layers map closely to the phases of the IoT data lifecycle with lookup/orchestration considered to be an added process that is not strictly a part of the data lifecycle. The *"Things" Layer* encompasses IoT sensors and smart objects (data production objects), as well as modules for in-network processing and data collection/real-time aggregation (processing,

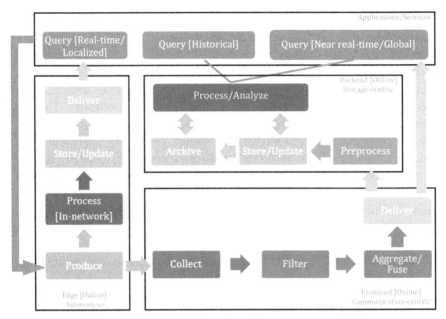

Figure 3.26   IoT data lifecycle and data management [73].

aggregation). The *Communication Layer* provides support for transmission of requests, queries, data, and results (collection and delivery). The *Data/Sources twin layers* respectively handle the discovery and cataloguing of data sources and the storage and indexing of collected data (data storage/archival). The Data Layer also handles data and query processing for local, autonomous data repository sites (filtering, pre-processing, processing).

The *Federation Layer* provides the abstraction and integration of data repositories that is necessary for global query/analysis requests, using meta-data stored in the Data Sources layer to support real-time integration of sources as well as location-centric requests (pre-processing, integration, fusion). The *Query Layer* handles the details of query processing and optimization in cooperation with the *Federation Layer* as well as the complementary *Transactions Layer* (processing, delivery). The Query Layer includes the *Aggregation Sub-Layer*, which handles the aggregation and fusion queries that involve an array of data sources/sites (aggregation/fusion). The *Application/ Analysis Layer* is the requester of data/analysis needs and the consumer of data and analysis results. The layers of the proposed IoT data management framework and their respective functional modules are illustrated in Figure 3.27 [73].

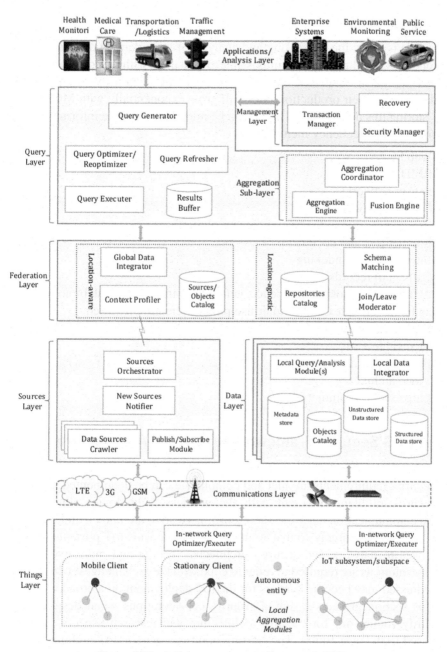

Figure 3.27 IoT data management framework [73].

3.7.1 Smart Data

Smart data is about the processing and analysis of large data repositories, so disproportionately large that it is impossible to treat them with the conventional tools of analytical databases. The machines generate data a lot faster than people can, and their production rates will grow exponentially with Moore's Law. Storing this data is cheap, and it can be mined for valuable information. Examples of this tendency include:

- Web logs
- RFID
- Sensor networks
- Social networks
- Social data (due to the Social data revolution)
- Internet text and documents
- Internet search indexing
- Call detail records
- Astronomy, atmospheric science, genomics, biogeochemical, biological, and other complex and/or interdisciplinary scientific research
- Military surveillance
- Medical records
- Photography archives
- Video archives
- Large scale e-commerce

3.8 A QoS Security Framework for the IoT Architecture

A Quality of Service (QoS) security framework would first and foremost mean that security requirements are met and compliance can be documented.

Security problems inherent with the wireless technologies (Internet, mobile communication networks, and sensor networks) are known and many of them addressed largely so that solutions are on the way. IoT presents new challenges to network and security architects. Specific and more evolved security solutions are required in order to cope with these challenges, which if not addressed may become barriers for the IoT deployment on a broad scale.

This section presents essential security considerations when designing a security framework for the IoT architecture and research aspects to be addressed in the near future. The starting point is a generic IoT architecture integrating physical objects communicating with each other and structured in several layers, suitable for resources-constrained devices. Security aspects are addressed tailored to constraints of IoT scenarios and characteristics of IoT devices.

The basic components of a QoS security framework are identified, addressing both traditional security problems of communication networks and specific IoT threats. Larger space is dedicated to authentication and access control as important parts of any security chain, and vital for many scenarios in the IP-based IoT. They have their own specificity and have been the focus of recent standardization and certification efforts.

3.8.1 End-to-End Security. The Decentralized Approach.

Large-scale applications and services based on the IoT are increasingly vulnerable to disruption from attack or information theft. Vulnerability is the opportunity for a threat to cause loss and a threat is any potential danger to a resource, originating from anything and/or anyone that has the potential to cause a threat. Common IoT threats are presented in [72] together with requirements to make the IoT secure, involving several technological areas. The thread that is common through all these is the need for end-to-end security.

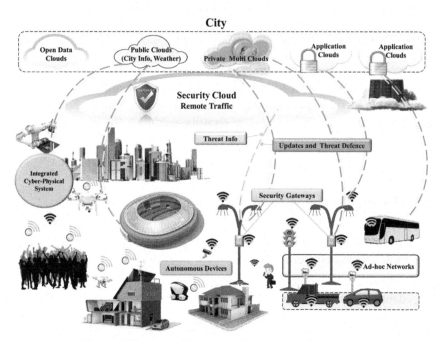

Figure 3.28 Smart City – Multi-layer security framework.

In order to fulfil the end-to-end security principles and IoT inherent requirements, a distributed approach seems to be the most suitable. With this approach, objects are becoming more intelligent, capable of making their own authorization decisions. The adoption of fine-grained authorization mechanisms allows for more flexible resources control and enables tolerance when fronting unknown-risks. In addition, IP security protocol variants for the IoT with public-key-based cryptographic primitives in their protocol design, such as Datagram TLS (DTLS), the HIP Diet EXchange (DEX), and minimal IKEv2, can fulfil the requirements of IoT regarding scalability and interoperability. End-to-end authentication, integrity confidentiality and privacy, are essential.

Important to keep in mind is that all the technologies must be tailored to the constraints of IoT scenarios and characteristics of IoT devices, including limited memory, compute resources, local security, backup connectivity. Thus the employed technologies must reduce the need for expensive cryptographic operations, prevent DoS attacks targeting the security mechanism, improve tolerance to attacks, etc.

3.8.2 Standardization. Certification. Interoperability.

Standardization and certification activities play an important role in securing the IoT, both in terms of enhancing interoperability of IoT devices and adoption of security solutions by the industry. Many of the security solutions are proprietary making it difficult for the IoT devices to communicate with each other in an interoperable manner and in formulation a common and sound security vision in order to standardize security solutions for the IoT. The efforts in the Internet Engineering Task Force (IETF) are making progress exactly in this direction.

3.8.3 Components of a QoS Security Framework

A Quality of Service (QoS) security framework would first and foremost mean that security requirements are met and compliance can be documented.

The basic components of a QoS security framework are:

- Authentication. Implements authentication of users and devices, including identity management in order to ensure authentication, accountability and privacy.
- Authorization. Implements access control on devices and services, in order to ensure data confidentiality and integrity.

- Network. Implements protocols to route and transport control, management and traffic securely over the infrastructure, thus ensuring communication confidentiality and integrity.
- Trust management. Implements remote control, over-the-air update, logging, analytics, in order to document compliance with security requirements and other security related regulations and standards.

3.8.3.1 Authentication

At the core of the security framework for IoT architecture is the authentication component, used to provide and verify the identify information of IoT objects.

According to ISO/IEC 27002, authentication is the act of establishing, or confirming something (or someone) as authentic, i.e., that claims made by, or about the thing are true. Thus, an authentication relationship is initiated based on the identity of the IoT device, whenever the device needs access to the IoT infrastructure.

Some of the traditional authentication technologies in wireless networks, including lightweight public key-based authentication technology, pre-shared key authentication technology, random key pre-distribution authentication technology, the certification based on auxiliary information, the certification based on one-way hash function, can be employed.

However, most of the traditional authentication mechanisms are based on human credentials, such as username and password, token or biometrics, roles in the organizations, etc. For the IoT objects the identity information is different, first and foremost because the process does not involve human intervention. Such information includes RFID, X.509 certificates, MAC address or any other unique hardware based information. However, many devices may have limited memory to store certificates or CPU power to execute the validation operations inherent to such certificates. Near future research much therefore address other credential types.

Another challenge deriving from the IoT devices being usually unattended, is the fact that the equipment is accessible to attacks, targeting exactly the security mechanisms. Accessing the IoT infrastructure with hacked/illegal equipment can create serious damages for the users, such as conflict of interests in addition to the network security issues.

3.8.3.2 Authorization

Authorization is the next component in the security chain, building upon the information provided by the authentication component. Both authentication and authorization must be in place in order to establish a secure relationship between IoT devices to exchange appropriate information.

According to ISO/IEC 27002, authorization is the process of controlling access and rights to resources.

The state of art and practice for policy mechanisms to manage and control access to consumer and enterprise networks is well advanced so it would only be natural to adopt them for the IoT. The most common scenario is that for a user to have the privileges to access a resource, the user must satisfy certain conditions, such as being assigned certain roles, belonging to certain specific groups, etc.

However, it is clear based on the type of identity information delivered by the authentication component that the traditional role-based access control mechanism (RBAC) is no longer the focus. In the IoT world, the attributes of a node or an IoT object make more sense, so that a fine-grained mechanism such as the attribute-based access control (ABAC) is more suitable.

Although being standard technologies, RBAC and ABAC cannot be applied straightforward to IoT. The challenge with the IoT is that very often there are many different contexts around an IoT identity, so that a centralized solution would not be feasible. The decisions must be made by the IoT objects able to capture local information. Authorization within a central entity would also impact on the scalability of the solution.

A multi context-aware authorization mechanism is necessary. Environment conditions which are captured locally by IoT end-devices may also come into the picture. The authorization component becomes more complex because at any point in time during the authorization process it should be clear: who is requesting the access, who is granting the access, what specific access is being requested, what is the access scope, when is the access requested and granted/denied, what is the access's duration?

A key challenge is therefore the IoT devices capability to capture security-relevant contextual information, such as time, location, state of the environment, etc., and use it to make access decision, when the access requests are issued. More research is needed in this direction.

3.8.3.3 Network

This component encompasses the elements that route and transport endpoint traffic securely over the infrastructure, whether control, management or actual data traffic. There are already established protocols and mechanisms to secure the network infrastructure and affect policy that are well suited to the IoT.

3.8.3.4 Trust Management

This is the component responsible for remote control, over-the-air update, logging of all security-related activities in the IoT environment, producing statistics and document compliance with all security requirements.

As IoT-scale applications and services will scale over multiple administrative domains and involve multiple ownership regimes, there is a need for a trust framework to enable the users of the system to have confidence that the information and services being exchanged can indeed be relied upon. It needs to be able to deal with humans and machines as users, i.e. it needs to convey trust to humans and needs to be robust enough to be used by machines without denial of service.

Trust can only be achieved by building continuous compliance into the IoT infrastructure. By this we mean that embedding techniques into the IoT devices that allow at any point in time to prove (as opposed to only claiming) that the IoT environment complies with the security requirements, the ever-changing laws and regulations related to security and other interoperability requirements inherent to the modern, more complex IoT environments.

3.9 Discussion

In the future the enterprises will make extensive use of IoT technology, and there will be a wide range of products sold into various markets, such as advanced medical devices; factory automation sensors and applications in industrial robotics; sensor motes for increased agricultural yield; and automotive sensors and infrastructure integrity monitoring systems for diverse areas, such as road and railway transportation, water distribution and electrical transmission. By 2020, component costs will have come down to the point that connectivity will become a standard feature, even for processors costing less than $1. This opens up the possibility of connecting just about anything, from the very simple to the very complex, to offer remote control, monitoring and sensing and it is expected that the variety of devices offered to explode [84].

The economic value added at the European and global level is significant across sectors in 2020. The IoT applications are still implemented by the different industrial verticals with a high adoption in manufacturing, healthcare and home/buildings.

IoT will also facilitate new business models based on the real-time data acquired by billions of sensor nodes. This will push for development of advances sensor, nanoelectronics, computing, and network and cloud technologies and will lead to value creation in utilities, energy, smart building technology, transportation and agriculture.

The IoT's paradigm is based around the idea of connecting things to each other, so it's essential create technology ecosystems and work with other companies that excel at creating IoT devices, gateways, communication/cloud computing platforms, services and applications. As the number of telecommunications providers, device manufacturers, consulting firms, and business software companies supplying IoT services grows, it's easier for enterprises to find the right providers with whom to partner. In order to address the totality of interrelated technologies the IoT technology ecosystem is essential and the enabling technologies will have different roles such as components, products/applications, and support and infrastructure in these ecosystems. The technologies will interact through these roles and impact the IoT technological deployment [50].

IoT ecosystems offer solutions comprising a large system beyond a platform and solve important technical challenges in the different verticals and across verticals. These IoT technology ecosystems are instrumental for the deployment of large pilots and can easily be connected to or build upon the core IoT solutions for different applications in order to expand the system of use and allow new and even unanticipated IoT end uses.

The IoT architecture needs to consider key scenarios, to design for common problems, to appreciate the long term consequences of key decisions in such a way that builds a solid foundation for developing IoT applications, based on specific scenarios and requirements. This is essential for both developing the IoT ecosystems and deploying successfully large IoT application pilots.

If the IoT architecture is not good enough and the software developed is unstable, the development is unable to support existing or future business requirements, and it is difficult to deploy or manage it in a large IoT pilot environment.

One challenge is exchanging the data from and among the things/objects in an interoperable format. This requires creating systems that cross vertical silos and harvest the data across domains, which unleashes useful IoT applications that are user centric, context aware, and are able to create new services by communication across those verticals.

These exchange and processing capabilities are an intrinsic part of the IoT concept and they can be applied to applications in areas such as the Internet of Energy (IoE), the Internet of Lighting (IoL), the Internet of Buildings (IoB), and, in a city context, the Internet of Vehicles (IoV).

The final aim is to create a city-centric ecosystem comprising state-of-the-art and viable technologies which apply the IoT, IoE and IoV concepts to increase the city efficiency by enabling unobtrusive, adaptable and highly usable services at the network-edge, gateway and cloud levels. In this context stimulating the creation of IoT ecosystems (comprising of stakeholders representing the IoT application value-chain: components, chips, sensors, actuators, embedded processing and communication, system integration, middleware, architecture design, software, security, service provision, usage, test, etc.), integrating the future generations of applications, devices, embedded systems and network technologies and other evolving ICT advances, based on open platforms and standardised identifiers, protocols and architectures is of paramount importance. In addition the deployment of IoT Large Scale Pilots to promote the market emergence of IoT and overcome the fragmentation of vertically oriented closed systems, architectures and application areas that address challenges in different application areas by bringing together the technology supply and the application demand sides in real-life settings is the next important step to demonstrate and validate the technology in real environments [50].

Acknowledgments

The IoT European Research Cluster – European Research Cluster on the Internet of Things (IERC) maintains its Strategic Research and Innovation Agenda (SRIA), taking into account its experiences and the results from the on-going exchange among European and international experts.

The present document builds on the 2010, 2011, 2012, 2013 and 2014 Strategic Research and Innovation Agendas and presents the research fields and an updated roadmap on future R&D from 2015 to 2020 and beyond 2020.

The IoT European Research Cluster SRIA is part of a continuous IoT community dialogue supported by the European Commission (EC) DG Connect – Communications Networks, Content and Technology, E1 – Network technologies Unit for the European and international IoT stakeholders. The result is a lively document that is updated every year with expert feedback from on-going and future projects financed by the EC. Many colleagues have assisted over the last few years with their views on the Internet of Things Strategic Research and Innovation agenda document. Their contributions are gratefully acknowledged.

Internet of Things Timelines

Table 3.2 Future Technological Developments

Development	2015–2020	Beyond 2020
Identification Technology	• Identity management • Open framework for the IoT • Soft Identities • Semantics • Privacy awareness	• "Thing/Object DNA" identifier • Context aware identification
Internet of Things Architecture Technology	• Network of networks architectures • IoT reference architecture developments • IoT reference architecture standardization • Adaptive, context based architectures • Self-* properties	• Cognitive architectures • Experimental architectures
Internet of Things Infrastructure	• Cross domain application deployment • Integrated IoT infrastructures • Multi-application infrastructures • Multi provider infrastructures	• Global, general purpose IoT infrastructures • Global discovery mechanism
Internet of Things Applications	• Configurable IoT devices • IoT in food/water production and tracing • IoT in manufacturing industry • IoT in industrial lifelong service and maintenance • IoT device with strong processing and analytics capabilities • Application capable of handling heterogeneous high capability data collection and processing infrastructures	• IoT information open market • Autonomous Vehicles • Internet of Buildings • Internet of Energy • Internet of Vehicles • Internet of Lighting

Communication Technology	• Wide spectrum and spectrum aware protocols • Ultra low power chip sets • On chip antennas • Millimetre wave single chips • Ultra low power single chip radios • Ultra low power system on chip	• Unified protocol over wide spectrum • Multi-functional reconfigurable chips
Network Technology	• Network context awareness • Self-aware and self-organizing networks • Sensor network location transparency • IPv6- enabled scalability	• Network cognition • Self-learning, self-repairing networks • Ubiquitous IPv6-based IoT deployment
Software and algorithms	• Goal oriented software • Distributed intelligence, problem solving • Things-to-Things collaboration environments • IoT complex data analysis • IoT intelligent data visualization • Hybrid IoT and industrial automation systems	• User oriented software • The invisible IoT • Easy-to-deploy IoT SW • Things-to-Humans collaboration • IoT 4 All • User-centric IoT • Nano-technology and new materials
Hardware	• Smart sensors (bio-chemical) • More sensors and actuators (tiny sensors) • Sensor integration with NFC • Home printable RFID tags	
Data and Signal Processing Technology	• Context aware data processing and data responses • Energy, frequency spectrum aware data processing	• Cognitive processing and optimisation

(*Continued*)

Table 3.2   Continued

Development	2015–2020	Beyond 2020
Discovery and Search Engine Technologies	• Automatic route tagging and identification management centres • Semantic discovery of sensors and sensor data	• Cognitive search engines • Autonomous search engines
Power and Energy Storage Technologies	• Energy harvesting (biological, chemical, induction) • Power generation in harsh environments • Energy recycling • Long range wireless power • Wireless power	• Biodegradable batteries • Nano-power processing unit
Security, Privacy & Trust Technologies	• User centric context-aware privacy and privacy policies • Privacy aware data processing • Security and privacy profiles selection based on security and privacy needs • Privacy needs automatic evaluation • Context centric security • Homomorphic Encryption • Searchable Encryption • Protection mechanisms for IoT DoS/DdoS attacks	• Self-adaptive security mechanisms and protocols • Self-managed secure IoT

Material Technology	• SiC, GaN • Improved/new semiconductor manufacturing processes/technologies for higher temperature ranges	• Diamond • Graphene
Interoperability	• Optimized and market proof interoperability approaches used • Interoperability under stress as market grows • Cost of interoperability reduced • Several successful certification programmes in place	• Automated self-adaptable and agile interoperability
Standardisation	• IoT standardization refinement • M2M standardization as part of IoT standardisation • Standards for cross interoperability with heterogeneous networks • IoT data and information sharing	• Standards for autonomic communication protocols

Table 3.3 Internet of Things Research Needs

Research Needs	2015–2020	Beyond 2020
Identification Technology	• Convergence of IP and IDs and addressing scheme • Unique ID • Multiple IDs for specific cases • Extend the ID concept (more than ID number) • Electro Magnetic Identification – EMID	• Multi methods – one ID
IoT Architecture	• Internet (Internet of Things) (global scale applications, global interoperability, many trillions of things)	
Internet of Things Infrastructure	• Application domain-independent abstractions & functionality • Cross-domain integration and management • Large-scale deployment of infrastructure • Context-aware adaptation of operation	• Self-management and configuration
Internet of Things Applications	• IoT information open market • Standardization of APIs • IoT device with strong processing and analytics capabilities • Ad-hoc deployable and configurable networks for industrial use • Mobile IoT applications for IoT industrial operation and service/maintenance • Fully integrated and interacting IoT applications for industrial use	• Building and deployment of public IoT infrastructure with open APIs and underlying business models • Mobile applications with bio-IoT-human interaction

SOA Software Services for IoT	• Quality of Information and IoT service reliability • Highly distributed IoT processes • Semi-automatic process analysis and distribution	
Internet of Things Architecture Technology	• Code in tags to be executed in the tag or in trusted readers • Global applications • Adaptive coverage • Universal authentication of objects • Graceful recovery of tags following power loss • More memory • Less energy consumption • 3-D real time location/position embedded systems	• Intelligent and collaborative functions • Object intelligence • Context awareness • Cooperative position cyber-physical systems
Communication Technology	• Longer range (higher frequencies – tenths of GHz) • Protocols for interoperability • On chip networks and multi standard RF architectures • Multi-protocol chips • Gateway convergence • Hybrid network technologies convergence • 5G developments • Collision-resistant algorithms • Plug and play tags • Self-repairing tags	• Self-configuring, protocol seamless networks
		• Fully autonomous IoT devices

(Continued)

Table 3.3   Continued

Research Needs	2015–2020	Beyond 2020
Network Technology	• Grid/Cloud network • Software defined networks • Service based network • Multi authentication • Integrated/universal authentication • Brokering of data through market mechanisms • Scalability enablers • IPv6-based networks for smart cities	• Need based network • Internet of Everything • Robust security based on a combination of ID metrics • Autonomous systems for nonstop information technology service • Global European IPv6-based Internet of Everything • Self-generating "molecular" software • Context aware software

Software and algorithms	• Self-management and control
	• Micro operating systems
	• Context aware business event generation
	• Interoperable ontologies of business events
	• Scalable autonomous software
	• Evolving software
	• Self-reusable software
	• Autonomous things:
	  • Self-configurable
	  • Self-healing
	  • Self-management
	• Platform for object intelligence	

Hardware Devices	• Polymer based memory • Ultra low power EPROM/FRAM • Molecular sensors • Autonomous circuits • Transparent displays • Interacting tags • Collaborative tags • Heterogeneous integration • Self-powering sensors	• Biodegradable circuits • Autonomous "bee" type devices

Hardware Systems, Circuits and Architectures	• Low cost modular devices • Ultra low power circuits • Electronic paper • Nano power processing units • Silent Tags • Biodegradable antennae • Multi-protocol front ends • Ultra low cost chips with security • Collision free air to air protocol • Minimum energy protocols • Multi-band, multi-mode wireless sensor architectures implementations • Adaptive architectures • Reconfigurable wireless systems • Changing and adapting functionalities to the environments • Micro readers with multi standard protocols for reading sensor and actuator data • Distributed memory and processing • Low cost modular devices • Protocols correct by construction	• Heterogeneous architectures • "Fluid" systems, continuously changing and adapting
Data and Signal Processing Technology	• Common sensor ontologies (cross domain) • Distributed energy efficient data processing • Autonomous computing • Tera scale computing • Micro servers • Multi-functional gateways	• Cognitive computing • Cognitive, software-defined gateways

(Continued)

Table 3.3 Continued

Research Needs	2015–2020	Beyond 2020
Discovery and Search Engine Technologies	• Scalable Discovery services for connecting things with services while respecting security, privacy and confidentiality • "Search Engine" for Things • IoT Browser • Multiple identities per object • On demand service discovery/integration • Universal authentication	• Cognitive registries
Power and Energy Storage Technologies	• Paper based batteries • Wireless power everywhere, anytime • Photovoltaic cells everywhere • Energy harvesting • Power generation for harsh environments	• Biodegradable batteries
Interoperability	• Dynamic and adaptable interoperability for technical and semantic areas • Open platform for IoT validation	• Self-adaptable and agile interoperability approaches
Security, Privacy & Trust Technologies	• Low cost, secure and high performance identification/authentication devices • Access control and accounting schemes for IoT • General attack detection and recovery/resilience for IoT • Cyber Security Situation Awareness for IoT • Context based security activation algorithms • Service triggered security • Context-aware devices	• Cognitive security systems • Self-managed secure IoT • Decentralised approaches to privacy by information localisation

	• Object intelligence • Decentralised self-configuring methods for trust establishment • Novel methods to assess trust in people, devices and data • Location privacy preservation • Personal information protection from inference and observation • Trust Negotiation	
Governance (legal aspects)	• Legal framework for transparency of IoT bodies and organizations • Privacy knowledge base and development privacy standards	• Adoption of clear European norms/standards regarding Privacy and Security for IoT
Economic	• Business cases and value chains for IoT • Emergence of IoT in different industrial sectors	• Integrated platforms
Material Technology	• Carbon nanotube • Conducting Polymers and semiconducting polymers and molecules • Modular manufacturing techniques	• Graphene

List of Contributors

Abdur Rahim Biswas, IT, create-net, iCore
Alessandro Bassi, FR, Bassi Consulting, IoT-A
Ali Rezafard, IE, Afilias, EPCglobal Data Discovery JRG
Amine Houyou, DE, SIEMENS, IoT@Work
Antonio Skarmeta, SP, University of Murcia, IoT6
Carlos Agostinho, PT, UNINOVA
Carlo Maria Medaglia, IT, University of Rome 'Sapienza', IoT-A
César Viho, FR, Probe-IT
Claudio Pastrone, IT, ISMB, ebbits, ALMANAC
Daniel Thiemert, UK, University of Reading, HYDRA
David Simplot-Ryl, FR, INRIA/ERCIM, ASPIRE
Elias Tragos, GR, FORTH, RERUM
Eric Mercier, FR, CEA-Leti
Erik Berg, NO, Telenor, IoT-I
Francesco Sottile, IT, ISMB, BUTLER
Franck Le Gall, FR, Inno, PROBE-IT, BUTLER
François Carrez, GB, IoT-I
Frederic Thiesse, CH, University of St. Gallen, Auto-ID Lab
Friedbert Berens, LU, FB Consulting S.à r.l, BUTLER
Gary Steri, IT, EC, JRC
Gianmarco Baldini, IT, EC, JRC
Giuseppe Abreu, DE, Jacobs University Bremen, BUTLER
Ghislain Despesse, FR, CEA-Leti
Harald Sundmaeker, DE, ATB GmbH, SmartAgriFood, CuteLoop
Henri Barthel, BE, GS1 Global
Igor Nai Fovino, IT, EC, JRC
Jan Höller, SE, EAB
Jens-Matthias Bohli, DE, NEC
John Soldatos, GR, Athens Information Technology, ASPIRE, OpenIoT
Jose-Antonio, Jimenez Holgado, ES, TID
Klaus Moessner, UK, UNIS, IoT.est
Kostas Kalaboukas, GR, SingularLogic, EURIDICE
Latif Ladid, LU, UL, IPv6 Forum
Levent Gürgen, FR, CEA-Leti
Luis Muñoz, ES, Universidad De Cantabria
Marco Carugi, IT, ITU-T, ZTE
Marilyn Arndt, FR, Orange

Mario Hoffmann, DE, Fraunhofer-Institute SIT, HYDRA

Markus Eisenhauer, DE, Fraunhofer-FIT, HYDRA, ebbits

Markus Gruber, DE, ALUD

Martin Bauer, DE, NEC, IoT-A

Martin Serrano, IE, OpenIoT, NUI Galway, Insight Centre, OpenIoT, VITAL

Maurizio Spirito, IT, Istituto Superiore Mario Boella, ebbits, ALMANAC

Maarten Botterman, NL, GNKS, SMART-ACTION

Nicolaie L. Fantana, DE, ABB AG

Nikos Kefalakis, GR, Athens Information Technology, OpenIoT

Paolo Medagliani, FR, Thales Communications & Security, CALYPSO

Payam Barnaghi, UK, UNIS, IoT.est

Philippe Cousin, FR, easy global market, PROBE-IT

Raffaele Giaffreda, IT, CNET, iCore

Ricardo Neisse, IT, EC, JRC

Richard Egan, UK, TRT

Rolf Weber, CH, UZH

Sébastien Boissseau, FR, CEA-Leti

Sébastien Ziegler, CH, Mandat International, IoT6

Sergio Gusmeroli, IT, TXT e-solutions,

Stefan Fisher, DE, UZL

Stefano Severi, DE, Jacobs University Bremen, BUTLER

Srdjan Krco, RS, DunavNET, IoT-I, SOCIOTAL

Sönke Nommensen, DE, UZL, SmartSantander

Trevor Peirce, BE, CASAGRAS2

Veronica Gutierrez Polidura, ES, Universidad De Cantabria

Vincent Berg, FR, CEA-Leti

Vlasios Tsiatsis, SE, EAB

Wolfgang König, DE, ALUD

Wolfgang Templ, DE, ALUD

Contributing Projects and Initiatives

ASPIRE, BRIDGE, CASCADAS, CONFIDENCE, CuteLoop, DACAR, ebbits, ARTEMIS, ENIAC, EPoSS, EU-IFM, EURIDICE, GRIFS, HYDRA, IMS2020, Indisputable Key, iSURF, LEAPFROG, PEARS Feasibility, PrimeLife, RACE networkRFID, SMART, StoLPaN, SToP, TraSer, WALTER, IoT-A, IoT@Work, ELLIOT, SPRINT, NEFFICS, IoT-I, CASAGRAS2, eDiana, OpenIoT, IoT6, iCore PROBE-IT, BUTLER, IoT-est, SmartAgri-Food, ALMANAC, CITYPULSE, COSMOS, CLOUT, RERUM, SMARTIE, SMART-ACTION, SOCIOTAL, VITAL.

List of Abbreviations and Acronyms

Acronym	Meaning
3GPP	3rd Generation Partnership Project
AAL	Ambient Assisted Living
ACID	Atomicity, Consistency, Isolation, Durability
ACL	Access Control List
AMR	Automatic Meter Reading Technology
API	Application Programming Interface
ARM	Architecture Reference Model
AWARENESS	EU FP7 coordination action
	Self-Awareness in Autonomic Systems
BACnet	Communications protocol for building automation and control networks
BAN	Body Area Network
BDI	Belief-Desire-Intention architecture or approach
Bluetooth	Proprietary short range open wireless technology standard
BPM	Business process modelling
BPMN	Business Process Model and Notation
BUTLER	EU FP7 research project
	uBiquitous, secUre inTernet of things with Location and contExt-awaReness
CAGR	Compound annual growth rate
CE	Council of Europe
CEN	Comité Européen de Normalisation
CENELEC	Comité Européen de Normalisation Électrotechnique
CEO	Chief executive officer
CEP	Complex Event Processing
CSS	Chirp Spread Spectrum
D1.3	Deliverable 1.3
DATEX-II	Standard for data exchange involving traffic centres
DCA	Data Collection and Analysis
DNS	Domain Name System
DoS/DDOS	Denial of service attack
	Distributed denial of service attack

EC	European Commission
eCall	eCall – eSafety Support
	A European Commission funded project,
	coordinated by ERTICO-ITS Europe
EDA	Event Driven Architecture
EH	Energy harvesting
EMF	Electromagnetic Field
ERTICO-ITS	Multi-sector, public/private partnership for
	intelligent transport systems and services for Europe
ESOs	European Standards Organisations
ESP	Event Stream Processing
ETSI	European Telecommunications Standards Institute
EU	European Union
Exabytes	10^{18} bytes
FI	Future Internet
FI PPP	Future Internet Public Private Partnership programme
FIA	Future Internet Assembly
FIS 2008	Future Internet Symposium 2008
F-ONS	Federated Object Naming Service
FP7	Framework Programme 7
FTP	File Transfer Protocol
GFC	Global Certification Forum
GreenTouch	Consortium of ICT research experts
GS1	Global Standards Organization
Hadoop	Project developing open-source software for reliable,
	scalable, distributed computing
IAB	Internet Architecture Board
IBM	International Business Machines Corporation
ICAC	International Conference on Autonomic Computing
ICANN	Internet Corporation for Assigned Name and Numbers
ICT	Information and Communication Technologies
iCore	EU research project
	Empowering IoT through cognitive technologies
IERC	European Research Cluster for the Internet of Things
IETF	Internet Engineering Task Force
INSPIRE	Infrastructure for Spatial Information in the European
	Community
IIoT	Industrial Internet of Things
IoB	Internet of Buildings

IoC	Internet of Cities
IoE	Internet of Energy
IoE	Internet of Everything
IoL	Internet of Lighting
IoM	Internet of Media
IoP	Internet of Persons, Internet of People
IoS	Internet of Services
IoT	Internet of Things
IoT6	EU FP7 research project
	Universal integration of the Internet of Things through an
	IPv6-based service oriented architecture enabling
	heterogeneous components interoperability
IoT-A	Internet of Things Architecture
IoT-est	EU ICT FP7 research project
	Internet of Things environment for service creation
	and testing
IoT-I	Internet of Things Initiative
IoV	Internet of Vehicles
IP	Internet Protocol
IPSO Alliance	Organization promoting the Internet Protocol
	(IP) for Smart Object communications
IPv6	Internet Protocol version 6
ISO 19136	Geographic information, Geography Mark-up Language,
	ISO Standard
IST	Intelligent Transportation System
KNX	Standardized, OSI-based network communications
	protocol for intelligent buildings
LNCS	Lecture Notes in Computer Science
LOD	Linked Open Data Cloud
LTE	Long Term Evolution
M2M	Machine to Machine
MAC	Media Access Control
	data communication protocol sub-layer
MAPE-K	Model for autonomic systems:
	Monitor, Analyse, Plan, Execute in interaction with a
	Knowledge base
makeSense	EU FP7 research project on
	Easy Programming of Integrated Wireless Sensors
MB	Megabyte

MIT	Massachusetts Institute of Technology
MPP	Massively parallel processing
NIEHS	National Institute of Environmental Health Sciences
NFC	Near Field Communication
NoSQL	not only SQL –
	a broad class of database management systems
OASIS	Organisation for the Advancement of Structured
	Information Standards
OEM	Original equipment manufacturer
OGC	Open Geospatial Consortium
OMG	Object Management Group
OpenIoT	EU FP7 research project
	Part of the Future Internet public private partnership
	Open source blueprint for large scale self-organizing
	cloud environments for IoT applications
Outsmart	EU project
	Provisioning of urban/regional smart services and
	business models enabled by the Future Internet
PAN	Personal Area Network
PET	Privacy Enhancing Technologies
Petabytes	10^{15} byte
PHY	Physical layer of the OSI model
PIPES	Public infrastructure for processing and exploring streams
PKI	Public key infrastructure
PPP	Public-private partnership
Probe-IT	EU ICT-FP7 research project
	Pursuing roadmaps and benchmarks for the Internet
	of Things
PSI	Public Sector Information
PV	Photo Voltaic
QoI	Quality of Information
RFID	Radio-frequency identification
SASO	IEEE international conferences on Self-Adaptive and
	Self-Organizing Systems
SDO	Standard Developing Organization
SEAMS	International Symposium on Software Engineering for
	Adaptive and Self-Managing Systems

SENSEI	EU FP7 research project
	Integrating the physical with the digital world of the network of the future
SIG	Special Interest Group
SLA	Service-level agreement/Software license agreement
SmartAgriFood	EU ICT FP7 research project
	Smart Food and Agribusiness: Future Internet for safe and healthy food from farm to fork
SmartSantander	EU ICT FP7 research project
	Future Internet research and experimentation
SOA	Service Oriented Approach
SON	Self-Organising Networks
SSW	Semantic Sensor Web
SRA	Strategic Research Agenda
SRIA	Strategic Research and Innovation Agenda
SRA2010	Strategic Research Agenda 2010
SWE	Sensor Web Enablement
TC	Technical Committee
TTCN-3	Testing and Test Control Notation version 3
USDL	Unified Service Description Language
UWB	Ultra-wideband
W3C	World Wide Web Consortium
WS& AN	Wireless sensor and actuator networks
WSN	Wireless sensor network
WS-BPEL	Web Services Business Process Execution Language
Zettabytes	10^{21} byte
ZigBee	Low-cost, low-power wireless mesh network standard based on IEEE 802.15.4

Bibliography

[1] NFC Forum, online at http://nfc-forum.org

[2] METIS, Mobile and wireless communications Enablers for the Twenty-twenty (2020) Information Society, online at https://www.metis2020.com/

[3] F. Schaich, B. Sayrac, and M. Schubert. On the Need for a New Air Interface for 5G, *IEEE COMSOC MMTC E-Letter*, Vol. 9, No. 5, September 2014, http://www.comsoc.org/~mmc

[4] Wemme, L., "NFC: Global Promise and Progress", NFC Forum, 22.01.2014, online at http://nfc-forum.org/wp-content/uploads/2014/01/Omnicard_Wemme_2014_website.pdf

[5] Bluetooth Special Interest Group, online at https://www.bluetooth.org/en-us/members/about-sig

[6] Bluetooth Developer Portal, online at https://developer.bluetooth.org/Pages/default.aspx

[7] Bluetooth, online at http://www.bluetooth.com

[8] ANT+, online at http://www.thisisant.com/

[9] ANT, "Message Protocol and Usage rev.5.0", online at http://www.thisisant.com/developer/resources/downloads#documents_tab

[10] ANT, "FIT2 Fitness Module Datasheet", online at http://www.thisisant.com/developer/resources/downloads#documents_tab

[11] Wi-Fi Alliance, online at http://www.wi-fi.org/

[12] Z-Wave alliance, online at http://www.z-wavealliance.org

[13] Pätz, C., "Smart lighting. How to develop Z-Wave Devices", EE|Times europe LEDLighting, 04.10.2012, online at http://www.ledlighting-eetimes.com/en/how-to-develop-z-wave-devices.html?cmp_id=71&news_id=222908151

[14] KNX, online at http://www.knx.org/knx-en/knx/association

[15] European Editors, "Using Ultra-Low-Power Sub-GHz Wireless for Self-Powered Smart-Home Networks", 12.05.2013, online at http://www.digikey.com/en-US/articles/techzone/2013/dec/using-ultra-low-power-sub-ghz-wireless-for-self-powered-smart-home-networks

[16] HART Communication Foundation, online at http://www.hartcomm.org

[17] Mouser Electronics, "Wireless Mesh Networking – Featured Wireless Mesh Networking Protocols", online at http://no.mouser.com/applications/wireless_mesh_networking_protocols/

[18] IETF, online at https://www.ietf.org

[19] Bormann, C., "6LoWPAN Roadmap and Implementation Guide", 6LoWPAN Working Group, April 2013, http://tools.ietf.org/html/draft-bormann-6lowpan-roadmap-04

[20] Shelby, Z. and Bormann, C., "6LoWPAN: The Wireless Embedded Internet", Wiley, Great Britain, ISBN 9780470747995, 2009, online at http://elektro.upi.edu/pustaka.elektro/Wireless%20Sensor%20Network/6LoWPAN.pdf

[21] WiMAX Forum, online at http://www.wimaxforum.org

[22] A. Passemard, "The Internet of Things Protocol stack – from sensors to business value", online at http://entrepreneurshiptalk.wordpress .com/2014/01/29/the-internet-of-thing-protocol-stack-from-sensors-to-business-value/

[23] EnOcean Alliance, online at http://www.enocean-alliance.org/en/pro file/

[24] EnOcean Wireless Standard, online at http://www.enocean.com

[25] EnOcean Alliance, "EnOcean Equipment Profiles (EEP)", Ver. 2.6, December 2013, online at http://www.enocean.com/en/home/

[26] DASH7 Alliance, online at http://www.dash7.org

[27] Maarten Weyn, "Dash7 Alliance Protocol Technical Presentation", December 2013, online at http://www.slideshare.net/MaartenWeyn1/ dash7-alliance-protocol-technical-presentation

[28] Visible Assets, Inc., "Rubee Technology", online at http://www.rubee .com/Techno/index.html

[29] Stevens, J., Weich, C., GilChrist, R., "RuBee (IEEE 1902.1) – The Physics Behind, Real-Time, High Security Wireless Asset Visibility Networks in Harsh Environments", online at http://www.rubee.com/ White-SEC/RuBee-Security-080610.pdf

[30] RuBee Hardware, online at http://www.rubee.com/page2/Hard/index .html

[31] Foster, A., "A Comparison Between DDS, AMQP, MQTT, JMS, REST and CoAP", Version 1.4, January 2014, online at http://www.prismtech.com/sites/default/files/documents/Messaging ComparsionJan2014USROW_vfinal.pdf

[32] Elkstein, M., "Learn REST: A tutorial", online at http://rest.elkstein.org

[33] Jaffey, T., "MQTT and CoAP IoT Protocols.pdf", September 2013, online at https://docs.google.com/document/d/1_kTNkl84o_yoC56dz FfkYHoHuepINP3nDNokycXINXI/edit?usp=sharing&pli=1

[34] Puzanov, O., "IoT Protocol Wars: MQTT vs COAP vs XMPP", online at http://www.iotprimer.com/2013/11/iot-protocol-wars-mqtt-vscoap-vs-xmpp.html

[35] Home Gateway Initiative (HGI), online at www.homegatewayinitiativ e.org

[36] Artemis IoE project, online at www.artemis-ioe.eu

[37] Casaleggio Associati, "The Evolution of Internet of Things", February 2011, online at http://www.casaleggio.it/pubblicazioni/Focus_inter net_of_things_v1.81%20-%20eng.pdf

[38] J. B., Kennedy, "When woman is boss, An interview with Nikola Tesla", in Colliers, January 30, 1926.

[39] M. Weiser, "The Computer for the 21st Century," *Scientific Am.*, Sept., 1991, pp. 94–104; reprinted in *IEEE Pervasive Computing*, Jan.–Mar. 2002, pp. 19–25."

[40] K. Ashton, "That 'Internet of Things' Thing", online at http://www.rfid journal.com/article/view/4986, June 2009

[41] N. Gershenfeld, "When Things Start to Think", Holt Paperbacks, New York, 2000.

[42] Raymond James & Associates, "The Internet of Things – A Study in Hype, Reality, Disruption, and Growth", online at http://sitic.org/wp-content/uploads/The-Internet-of-Things-A-Study-in-Hype-Reality-D isruption-and-Growth.pdf, January 2014.

[43] N. Gershenfeld, R. Krikorian and D. Cohen, *Scientific Am.*, Sept., 2004.

[44] World Economic Forum, "The Global Information Technology Report 2012 – Living in a Hyperconnected World" online at http://www3.weforum.org/docs/Global_IT_Report_2012.pdf

[45] "Key Enabling Technologies", Final Report of the HLG-KET, June 2011.

[46] G. Bovet, A. Ridi and J. Hennebert, "Toward Web Enhanced Building Automation System", in Eds. N. Bessis and C. Dobre – Big Data and Internet of Things: A Roadmap for Smart Environments, ISBN: 978-3-319-05028-7, Studies in Computational Intelligence, Volume 546, 2014 pp. 259–283, online at http://hal.archives-ouv ertes.fr/docs/00/97/35/10/PDF/BuildingsWoT.pdf

[47] International Technology Roadmap for Semiconductors, ITRS 2012 Update, online at http://www.itrs.net/Links/2012ITRS/2012Chapters/ 2012Overview.pdf

[48] W. Arden, M. Brillouët, P. Cogez, M. Graef, et al., "More than Moore" White Paper, online at http://www.itrs.net/Links/2010ITRS/IRC-ITRS-MtM-v2%203.pdf

[49] Frost & Sullivan "Mega Trends: Smart is the New Green" online at http://www.frost.com/prod/servlet/our-services-page.pag?mode= open&sid=230169625

[50] O. Vermesan. The IoT: a concept, a paradigm, and an open global network. *Telit2market International*, Issue 10, February 2015, pp. 120–122, online at http://www.telit2market.com/wp-content/

uploads/2015/02/telit2market_10_15_anniversary_edition.pdf, Accessed 29 May 2015.

[51] Market research group Canalys, online at http://www.canalys.com/

[52] E. Savitz, "Gartner: 10 Critical Tech Trends For The Next Five Years" online at http://www.forbes.com/sites/ericsavitz/2012/10/22/gartner-10-critical-tech-trends-for-the-next-five-years/

[53] E. Savitz, "Gartner: Top 10 Strategic Technology Trends For 2013" online at http://www.forbes.com/sites/ericsavitz/2012/10/23/gartner-top-10-strategic-technology-trends-for-2013/

[54] P. High "Gartner: Top 10 Strategic Technology Trends For 2014" online at http://www.forbes.com/sites/peterhigh/2013/10/14/gartner-top-10-strategic-technology-trends-for-2014/#

[55] Gartner's Top 10 Strategic Technology Trends for 2015, online at http://www.gartner.com/smarterwithgartner/gartners-top-10-strateg ic-technology-trends-for-2015/

[56] Platform INDUSTRIE 4.0 – Recommendations for implementing the strategic initiative INDUSTRIE 4.0, Final report of the Industrie 4.0 Working Group, online at, http://www.acatech.de/fileadmin/ user_upload/Baumstruktur_nach_Website/Acatech/root/de/Material_f uer_Sonderseiten/Industrie_4.0/Final_report__Industrie_4.0_accessible .pdf, 2013

[57] Industrial Internet of Things (IoT) Advisory Service, ARC Advisory Group, online at, http://www.arcweb.com/services/pages/industrial-internet-of-things-service.aspx

[58] rtSOA – A Data Driven, Real Time Service Oriented Architecture for Industrial Manufacturing, online at http://www-db.in.tum.de/research/projects/rtSOA/

[59] P. C. Evans and M. Annunziata, Industrial Internet: Pushing the Boundaries of Minds and Machines, General Electric Co., online at http://files.gereports.com/wp-content/uploads/2012/11/ge-industrial-internet-vision-paper.pdf

[60] Cisco, "Securely Integrating the Cyber and Physical Worlds", online at http://www.cisco.com/web/solutions/trends/tech-radar/securing-the-i ot.html

[61] NXT Cities, online at http://www.communicasia.com/wp-content/ themes/cmma2014/images/img-nxtcities-large.jpg

[62] NXT Enterprises, online at http://www.communicasia.com/wp-content/themes/cmma2014/images/img-nxtenterprise-large.jpg

[63] NXT Connect, online at http://www.communicasia.com/wp-content/themes/cmma2014/images/img-nxtconnect-large.jpg

[64] H. Bauer, F. Grawert, and S. Schink, Semiconductors for wireless communications: Growth engine of the industry, online at www.mckinsey.com/

[65] L. Fretwell and P. Schottmiller, Cisco Presentation, online at http://www.cisco.com/assets/events/i/nrf-Internet_of_Everything_Whats_the_Art_of_the_Possible_in_Retail.pdf, January 2014.

[66] ITU-T, Internet of Things Global Standards Initiative, http://www.itu.int/en/ITU-T/gsi/iot/Pages/default.aspx

[67] International Telecommunication Union – ITU-TY.2060 – (06/2012) – Next Generation Networks – Frameworks and functional architecture models – Overview of the Internet of things.

[68] IEEE-SA – Enabling Consumer Connectivity Through Consensus Building, online at http://standardsinsight.com/ieee_company_detail/consensus-building

[69] Mobile-Edge Computing – Introductory Technical White Paper, 2014, online at https://portal.etsi.org/Portals/0/TBpages/MEC/Docs/Mobile-edge_Computing_-_Introductory_Technical_White_Paper_V1%2018-09-14.pdf

[70] O. Vermesan, P. Friess, P. Guillemin, S. Gusmeroli, et al., "Internet of Things Strategic Research Agenda", Chapter 2 in Internet of Things – Global Technological and Societal Trends, River Publishers, 2011, ISBN 978-87-92329-67-7.

[71] O. Vermesan, P. Friess, P. Guillemin, H. Sundmaeker, et al., "Internet of Things Strategic Research and Innovation Agenda", Chapter 2 in Internet of Things – Converging Technologies for Smart Environments and Integrated Ecosystems, River Publishers, 2013, ISBN 978-87-92982-73-5.

[72] O. Vermesan, P. Friess, P. Guillemin, H. Sundmaeker, et al. Internet of Things Strategic Research and Innovation Agenda. O. Vermesan and P. Friess, Eds. *Internet of Things Applications – From Research and Innovation to Market Deployment*. Alborg, Denmark: The River Publishers, ISBN: 978-87-93102-94-1, 2014, pp. 7–142.

[73] M. Abu-Elkheir, M. Hayajneh and N. Abu Ali. Data Management for the Internet of Things: Design Primitives and Solution. *Sensors*, 13(11): 15582–15612, 2013.

[74] SmartSantander, EU FP7 project, Future Internet Research and Experimentation, online at http://www.smartsantander.eu/

[75] Introducing Fujisawa SST – A town sustainably evolving through living ideas, Panasonic, online at http://panasonic.net/es/fujisawasst/

[76] Foundational Solution of Smart City, online at http://br.fiberhomegrou p.com/pt/Enterprise/324/2282.aspx#1

[77] ISO 9646: "Conformance Testing Methodology and Framework".

[78] J. Bloem. InterOperability Testing in the Age of Cloud, Things and DevOps, online at https://www.sogeti.nl/updates/blogs/interoperabilit y-testing-age-cloud-things-and-devops

[79] Internet of Things Concept, online at http://xarxamobal.diba.cat/XGM SV/imatges/actualitat/iot.jpg

[80] H. Grindvoll, O. Vermesan, T. Crosbie, R. Bahr, et al., "A wireless sensor network for intelligent building energy management based on multi communication standards – a case study", ITcon Vol. 17, pg. 43–62, http://www.itcon.org/2012/3

[81] S. Toronidis. Better Operations, Better Operating Rooms, online at http://www.centrak.com/blog.aspx

[82] EU Research & Innovation, "Horizon 2020", The Framework Programme for Research and Innovation, online at http://ec.europa.eu/ research/horizon2020/index_en.cfm

[83] Digital Agenda for Europe, European Commission, Digital Agenda 2010–2020 for Europe, online at http://ec.europa.eu/information_socie ty/digital-agenda/index_en.htm

[84] Gartner, 2013, online at http://www.gartner.com/newsroom/id/ 2636073

[85] Beecham Research Limited. Towards Smart Farming: Agriculture Embracing the IoT Vision, online at http://www.beechamresearch .com/download.aspx?id=40

[86] E. Guizzo. How Google's Self-Driving Car Works. IEEE Spectrum, online at http://spectrum.ieee.org/automaton/robotics/artificial-intelligence/how-google-self-driving-car-works

[87] K. Karimi and G. Atkinson, "What the Internet of Things (IoT) Needs to Become a Reality", White Paper, 2013, online at http://www .freescale.com/files/32bit/doc/white_paper/INTOTHNGSWP.pdf

[88] Freescale vision chip makes self-driving cars a bit more ordinary, online at http://www.cnet.com/news/freescale-vision-chip-makes-self-driving-cars-a-bit-more-ordinary/

[89] R. E. Hall, "The Vision of A Smart City" presented at the 2nd International Life Extension Technology Workshop Paris, France September 28, 2000, online at http://www.crisismanagement.com.cn/

templates/blue/down_list/llzt_zhcs/The%20Vision%20of%20A%20S
mart%20City.pdf

[90] EU 2012. The ARTEMIS Embedded Computing Systems Initiative, October 2012 online at http://www.artemis-ju.eu/

[91] Foundations for Innovation in Cyber-Physical Systems, Workshop Report, NIST, 2013, online at http://www.nist.gov/el/upload/CPS-WorkshopReport-1-30-13-Final.pdf

[92] IERC – European Research Cluster on the Internet of Things, "Internet of Things – Pan European Research and Innovation Vision", October, 2011, online at, http://www.theinternetofthings.eu/sites/default/files/Rob%20van%20Kranenburg/IERC_IoT-Pan%20European%20Resea rch%20and%20Innovation%20Vision_2011.pdf

[93] O. Vermesan, P. Friess, G. Woysch, P. Guillemin, S. Gusmeroli, et al., "Europe's IoT Stategic Research Agenda 2012", Chapter 2 in The Internet of Things 2012 New Horizons, Halifax, UK, 2012, ISBN 978-0-9553707-9-3.

[94] SENSEI, EU FP7 project, *D1.4: Business models and Value Creation*, 2010, online at: http://www.ict-sensei.org

[95] IoT-I, Internet of Things Initiative, FP7 EU project, online at http://www.iot-i.eu

[96] Libelium, "50 Sensor Applications for a Smarter World", online at http://www.libelium.com/top_50_iot_sensor_applications_ranking#

[97] OUTSMART, FP7 EU project, part of the Future Internet Private Public Partnership, "OUTSMART – Provisioning of urban/regional smart services and business models enabled by the Future Internet", online at http://www.fi-ppp-outsmart.eu/en-uk/Pages/default.aspx

[98] BUTLER, FP7 EU project, online at http://www.iot-butler.eu/

[99] NXP Semiconductors N.V., "What's Next for Internet-Enabled Smart Lighting?", online at http://www.nxp.com/news/press-releases/2012/05/whats-next-for-internet-enabled-smart-lighting.html

[100] J. Formo, M. Gårdman, and J. Laaksolahti, "Internet of things marries social media", in *Proceedings of the 13th International Conference on MobileHCI,* ACM, New York, NY, USA, pp. 753–755, 2011.

[101] J. G. Breslin, S. Decker, and M. Hauswirth, et. al., "Integrating Social Networks and Sensor Networks", *W3C Workshop on the Future of Social Networking,* Barcelona, 15–16 January 2009.

[102] M. Kirkpatrick, "The Era of Location-as-Platform Has Arrived", *ReadWriteWeb,* January 25, 2010.

[103] F. Calabrese, K. Kloeckl, and C. Ratti (MIT), "WikiCity: Real-Time Location-Sensitive tools for the city", in *IEEE Pervasive Computing*, July–September 2007.

[104] Building smart communities, online at http://www.holyroodconnect .com/tag/smart-cities/

[105] Using Big Data to Create Smart Cities, online at http://informationstr ategyrsm.wordpress.com/2013/10/12/using-big-data-to-create-smart-cities/

[106] N. Maisonneuve, M. Stevens, M. E. Niessen, L. Steels, "NoiseTube: Measuring and mapping noise pollution with mobile phones", in *Information Technologies in Environmental Engineering (ITEE 2009)*, Proceedings of the 4th International ICSC SymposiumThessaloniki, Greece, May 28–29, 2009.

[107] J-S. Lee, B. Hoh, "Sell your experiences: a market mechanism based incentive for participatory sensing", *2010 IEEE International Conference on Pervasive Computing and Communications (PerCom)*, pp. 60–68, March 29, 2010, – April 2, 2010.

[108] R. Herring, A. Hofleitner, S. Amin, T. Nasr, A. Khalek, P. Abbeel, and A. Bayen, "Using Mobile Phones to Forecast Arterial Traffic Through Statistical Learning", *89th Transportation Research Board Annual Meeting,* Washington D.C., January 10–14, 2010.

[109] M. Kranz, L. Roalter, and F. Michahelles, "Things That Twitter: Social Networks and the Internet of Things", in *What can the Internet of Things do for the Citizen (CIoT) Workshop* at *The Eighth International Conference on Pervasive Computing (Pervasive 2010)*, Helsinki, Finland, May 2010.

[110] O. Vermesan, et al., "Internet of Energy – Connecting Energy Anywhere Anytime" in Advanced Microsystems for Automotive Applications 2011: Smart Systems for Electric, Safe and Networked Mobility, Springer, Berlin, 2011, ISBN 978-36-42213-80-9.

[111] W. Colitti, K. Steenhaut, and N. De Caro, "Integrating Wireless Sensor Networks with the Web," Extending the Internet to Low Power and Lossy Networks (IP+ SN 2011), 2011 online at http://hinrg.cs.jhu.edu/joomla/images/stories/IPSN_2011_koliti.pdf

[112] M. M. Hassan, B. Song, and E. Huh, "A framework of sensor-cloud integration opportunities and challenges", in *Proceedings of the 3rd International Conference on Ubiquitous Information Management and Communication,* ICUIMC 2009, Suwon, Korea, January 15–16, pp. 618–626, 2009.

[113] M. Yuriyama and T. Kushida, "Sensor-Cloud Infrastructure – Physical Sensor Management with Virtualized Sensors on Cloud Computing", NBiS 2010: 1–8.

[114] C. Bizer, T. Heath, K. Idehen, and T. Berners-Lee, "Linked Data on the Web", *Proceedings of the 17th International Conference on World Wide Web (WWW'08)*, New York, NY, USA, ACM, pp. 1265–1266, 2008.

[115] T. Heath and C. Bizer, "Linked Data: Evolving the Web into a Global Data Space", *Synthesis Lectures on the Semantic Web: Theory and Technology*, 1st edition. Morgan & Claypool, 1:1, 1–136, 2011.

[116] IBM, "An architectural blueprint for autonomic computing", IBM White paper. June 2005.

[117] Connected Devices for Smarter Home Environments, IBM Data Magazine, 2014, online at http://ibmdatamag.com/2014/04/connected-devices-for-smarter-home-environments/

[118] International Conference on Autonomic Computing http://www.autonomic-conference.org/

[119] IEEE International Conferences on Self-Adaptive and Self-Organizing Systems, http://www.saso-conference.org/

[120] International Symposium on Software Engineering for Adaptive and Self-Managing Systems, http://www.seams2012.cs.uvic.ca/

[121] Awareness project, Self-Awareness in Autonomic Systems http://www.aware-project.eu/

[122] M. C. Huebscher, J. A. McCann, "A survey of autonomic computing — degrees, models, and applications", *ACM Computing Surveys (CSUR)*, Volume 40 Issue 3, August 2008.

[123] A. S. Rao, M. P. Georgeff, "BDI Agents: From Theory to Practice", in *Proceedings of The First International Conference on Multi-agent Systems (ICMAS)*, 1995. pp. 312–319.

[124] G. Dimitrakopoulos, P. Demestichas, W. Koenig, *Future Network & Mobile Summit 2010 Conference Proceedings*.

[125] John Naisbit and Patricia Aburdene (1991), *Megatrends 2000*, Avon.

[126] D. C. Luckham, Event Processing for Business: Organizing the Real-Time Enterprise, John Wiley & Sons, 2012.

[127] T. Mitchell, *Machine Learning*, McGraw Hill, 1997.

[128] D. Estrin, "Participatory Sensing: Applications and Architecture, online at http://research.cens.ucla.edu/people/estrin/resources/conferences/2010-Estrin-participatory-sensing-mobisys.pdf O. Etzion, P. Niblett, *Event Processing in Action*, Manning, 2011.

[129] V. J. Hodgem, J. Austin, "A Survey of Outlier Detection Methodologies", *Artificial Intelligence Review*, 22 (2), pages 85–126, 2004.

[130] F. Angiulli, and C. Pizzuti, "Fast outlier detection in high dimensional spaces" in *Proc. European Conf. on Principles of Knowledge Discovery and Data Mining*, 2002.

[131] H. Fan, O. Zaïane, A. Foss, and J. Wu, "Nonparametric outlier detection for efficiently discovering top-n outliers from engineering data", in *Proc. Pacific-Asia Conf. on Knowledge Discovery and Data Mining (PAKDD)*, Singapore, 2006.

[132] A. Ghoting, S. Parthasarathy, and M. Otey, "Fast mining of distance-based outliers in high dimensional spaces", in *Proc. SIAM Int. Conf. on Data Mining (SDM)*, Bethesda, ML, 2006.

[133] G. Box, G. Jenkins, *Time series analysis: forecasting and control*, rev. ed., Oakland, California: Holden-Day, 1976.

[134] J. Hamilton, *Time Series Analysis*, Princeton Univ. Press, 1994.

[135] J. Durbin and S. J. Koopman, *Time Series Analysis by State Space Methods*, Oxford University Press, 2001.

[136] R. O. Duda, P. E. Hart, D. G. Stork, *Pattern Classification, 2nd Edition*, Wiley, 2000.

[137] C. M. Bishop, *Neural Networks for Pattern Recognition*, Oxford University Press, 1995.

[138] C. M. Bishop, *Pattern Recognition and Machine Learning*, Springer, 2006.

[139] M. J. Zaki, "Generating non-redundant association rules", *Proceedings of the sixth ACM SIGKDD international conference on Knowledge discovery and data mining*, 34–43, 2000.

[140] M. J. Zaki, M. Ogihara, "Theoretical foundations of association rules", *3rd ACM SIGMOD Workshop on Research Issues in Data Mining and Knowledge Discovery*, 1998.

[141] N. Pasquier, Y. Bastide, R. Taouil, L. Lakhal, "Discovering Frequent Closed Itemsets for Association Rules", *Proceedings of the 7th International Conference on Database Theory*, (398–416), 1999.

[142] C. M. Kuok, A. Fu, M. H. Wong, "Mining fuzzy association rules in databases", *SIGMOD Rec.* 27, 1 (March 1998), 41–46.

[143] T. Kohonen, *Self-Organizing Maps*, Springer, 2001.

[144] S.-H. Hamed, S. Reza, "TASOM: A New Time Adaptive Self-Organizing Map", *IEEE Transactions on Systems, Man, and Cybernetics—Part B: Cybernetics* 33 (2): 271–282, 2003.

[145] L.J.P. van der Maaten, G.E. Hinton, "Visualizing High-Dimensional Data Using t-SNE", *Journal of Machine Learning Research* 9(Nov): 2579–2605, 2008.

[146] I. Guyon, S. Gunn, M. Nikravesh, and L. Zadeh (Eds), *Feature Extraction, Foundations and Applications*, Springer, 2006.

[147] Y. Bengio, "Learning deep architectures for AI", *Foundations and Trends in Machine Learning*, 2(1):1–12, 2009.

[148] Y. Bengio, Y. LeCun, "Scaling learning algorithms towards AI", *Large Scale Kernel Machines*, MIT Press, 2007.

[149] B. Hammer, T. Villmann, "How to process uncertainty in machine learning?", *ESANN'2007 proceedings – European Symposium on Artificial Neural Networks*, Bruges (Belgium), 2007.

[150] J. Quinonero-Candela, C. Rasmussen, F. Sinz, O. Bousquet, and B. Schölkopf, "Evaluating Predictive Uncertainty Challenge", in *Machine Learning Challenges: Evaluating Predictive Uncertainty, Visual Object Classification, and Recognising Tectual Entailment*, First PASCAL Machine Learning Challenges Workshop (MLCW 2005), Springer, Berlin, Germany, 1–27, 2006.

[151] D. Koller and N. Friedman, *Probabilistic graphical models: principles and techniques*, MIT press, 2009.

[152] M. R. Endsley, "Measurement of situation awareness in dynamic systems", *Human Factors*, 37, 65–84, 1995.

[153] R. Fuller, *Neural Fuzzy System*, Åbo Akademi University, ESF Series A: 443, 1995, 249 pages. [ISBN 951-650-624-0, ISSN 0358-5654].

[154] S. Haykin, *Neural Networks: A Comprehensive Foundation, 2nd edn.*, Prentice-Hall, New York (1999).

[155] L. Rabiner, "A Tutorial on Hidden Markov Models and Selected Applications in Speech Recognition," *Proceedings of the IEEE*, vol. 77, no. 2, Feb. 1989.

[156] S.K. Murthy, "Automatic construction of decision trees from data: a multi-disciplinary survey", *Data Mining Knowledge Discovery*, 1998.

[157] ITU towards "IMT for 2020 and beyond."

[158] G. Macaigne. SIM-less networks, the new Eldorado of M2M and Internet of Things, online at http://www.inov360.com/blog/sim-less-networks-the-new-eldorado-of-m2m-and-internet-of-things/

[159] S. Haller and C. Magerkurth, "The Real-time Enterprise: IoT-enabled Business Processes", IETF IAB Workshop on Interconnecting Smart Objects with the Internet, March 2011.

[160] Open Geospatial Consortium, Geospatial and location standards, http://www.opengeospatial.org

[161] M. Botts, G. Percivall, C. Reed, and J. Davidson, "oGC Sensor Web Enablement: Overview and High Level Architecture", *The Open Geospatial Consortium*, 2008, online at http://portal.opengeospatial .org/files/?artifactid=25562

[162] W3C Semantic Sensor Network Incubator Group, Incubator Activity, online at http://www.w3.org/2005/Incubator/ssn/

[163] Logical Neighborhoods, Virtual Sensors and Actuators, online at http://logicalneighbor.sourceforge.net/vs.html

[164] K. M. Chandy and W. R. Schulte, "What is Event Driven Architecture (EDA) and Why Does it Matter?", 2007, online at http://complexevents.com/?p=212, (accessed on: 25.02.2008).

[165] D. Luckham, "What's the Difference Between ESP and CEP?", 2006, online at http://complexevents.com/?p=103, accessed on 15.12.2008.

[166] The CEP Blog, http://www.thecepblog.com/

[167] EnOcean – the Energy Harvesting Wirless Standard for Building Automation and Industrial Automation, online at http://www.enocean .com/en/radio-technology/

[168] IEEE Std 802.15.4^{TM}-2006, Wireless Medium Access Control (MAC) and Physical Layer (PHY) Specifications for Low-Rate Wireless Personal Area Networks (LR-WPANs), online at http://www.ieee802.org/15/pub/TG4.html

[169] Bluetooth Low Energy (LE) Technology Info Site, online at http://www.bluetooth.com/English/Products/Pages/low_energy.aspx

[170] The Official Bluetooth Technology Info Site, online at http://www .bluetooth.com/

[171] M-G. Di Benedetto and G. Giancola, *Understanding Ultra Wide Band Radio Fundamentals*, Prentice Hall, June 27, 2004.

[172] ISO, International Organization for Standardization (ISO), Identification cards – Contactless integrated circuit(s) cards – Vicinity cards, ISO/IEC 14443, 2003.

[173] N. Pletcher, S. Gambini, and J. Rabaey, "A 52 µW Wake-Up Receiver With 72 dBm Sensitivity Using an Uncertain-IF Architecture", in *IEEE Journal of Solid-State Circuits*, vol. 44, no1, January, pp. 269–280. 2009.

[174] A. Vouilloz, M. Declercq, and C. Dehollain, "A Low-Power CMOS Super-Regenerative Receiver at 1 GHz", in *IEEE Journal of Solid-State Circuits*, vol. 36, no3, March, pp. 440–451, 2001.

[175] J. Ryckaert, A. Geis, L. Bos, G. van der Plas, J. Craninckx, "A 6.1 GS/s 52.8 mW 43 dB DR 80 MHz Bandwidth 2.4 GHz RF Bandpass Σ-Δ ADC in 40 nm CMOS", in *IEEE Radio-Frequency Integrated Circuits Symposium*, 2010.

[176] L. Lolis, C. Bernier, M. Pelissier, D. Dallet, and J.-B. Bégueret, "Bandpass Sampling RX System Design Issues and Architecture Comparison for Low Power RF Standards", *IEEE ISCAS 2010*.

[177] D. Lachartre, "A 550 μW inductorless bandpass quantizer in 65 nm CMOS for 1.4-to-3 GHz digital RF receivers", *VLSI Circuits 2011*, pp. 166–167, 2011.

[178] S. Boisseau and G. Despesse, "Energy Harvesting, Wireless Sensor Networks & Opportunities for Industrial Applications", in *EETimes,* 27th Feb 2012, online at http://www.eetimes.com

[179] J.G. Koomey, S. Berard, M. Sanchez, and H. Wong, "Implications of Historical Trends in the Electrical Efficiency of Computing", in *IEEE Annals of the History of Computing*, vol. 33, no. 3, pp. 46–54, March 2011.

[180] eCall – eSafety Support, online at http://www.esafetysupport.org/en/ ecall_toolbox/european_commission/index.html

[181] European Commission, "Smart Grid Mandate, Standardization Mandate to European Standardisation Organisations (ESOs) to support European Smart Grid deployments", M/490 EN, Brussels 1st March, 2011.

[182] Global Certification Forum, online at http://www.globalcertificationfor um.org

[183] SENSEI, EU FP7 project, online at http://www.sensei-project.eu

[184] IoT-A, EU FP7 project, online at http://www.iot-a.eu

[185] IoT6, EU FP7 project, online at http://www.iot6.eu

[186] IoT@Work, EU FP7 project, online at https://www.iot-at-work.eu/

[187] Federated Object Naming Service, GS1, online at http://www.gs1.org/ gsmp/community/working_groups/gsmp#FONS

[188] Directive 2003/98/EC of the European Parliament and of the Council on the reuse of public sector information, 17 November 2003, online at http://ec.europa.eu/information_society/policy/psi/docs/pdfs/directi ve/psi_directive_en.pdf

[189] INSPIRE, EU FP7 project, – Infrastructure for Spatial Information in Europe, online at http://inspire.jrc.ec.europa.eu/

[190] H. van der Veer, A. Wiles, "Achieveing Technical Interoperability – the ETSI Approach", ETSI White Paper No.3, 3rd edition, April 2008,

http://www.etsi.org/images/files/ETSIWhitePapers/IOP%20whitepap er%20Edition%203%20final.pdf

[191] Ambient Assisted Living Roadmap, AALIANCE.

[192] Atmel AVR Xmega Micro Controllers, http://it.mouser.com/atmel_x mega/

[193] Worldwide Cellular M2M Modules Forecast, Beecham Research Ltd, August 2010.

[194] Future Internet Assembly Research Roadmap, FIA Research Roadmap Working Group, May 2011.

[195] D. Scholz-Reiter, M.-A. Isenberg, M. Teucke, H. Halfar, "An integrative approach on Autonomous Control and the Internet of Things", 2010.

[196] NIEHS on EMF, http://www.niehs.nih.gov/health/topics/agents/emf/

[197] R.H. Weber/R. Weber, "Internet of Things – Legal Perspectives", Springer, Berlin 2010.

[198] "The Global Wireless M2M Market", Berg Insight, 2010, http://www .berginsight.com/ReportPDF/ProductSheet/bi-gwm2m-ps.pdf

[199] M. Hatton, "Machine-to-Machine (M2M) communication in the Utilities Sector 2010–2020", Machina Research, July 2011.

[200] G. Masson, D. Morche, H. Jacquinot, and P. Vincent, "A 1 nJ/b 3.2–4.7 GHz UWB 50 Mpulses/s Double Quadrature Receiver for Communication and Localization", in *ESSCIRC 2010*.

4

Internet of Things Application Scenarios, Pilots and Innovation

Maurizio A. Spirito and Maria T. Delgado

ISMB, Italy

4.1 Introduction

One of the main challenges when developing a new technology or system is the understanding if and when they will be mature enough to reach the market. The field of the Internet of Things (IoT) is no exception; a lot of expectations have been generated regarding the uptake of IoT-based solutions due to the wide variety of applications domains in which IoT has the potential to change our lives [1].

Within the context of the Horizon 2020 research framework, which puts a stronger emphasis on innovation and exploitation of research results, the European Commission has recently stimulated the creation of multi-stakeholder ecosystems that are believed to be the key for the success of IoT in Europe. Now, in order to guarantee market take-up of the IoT solutions developed by EU-funded projects and initiatives, the research community needs to identify market gaps and possible ways to bridge them at the same time ensuring the sustainability of project outcomes beyond the projects lifetime.

This chapter aims to provide an insight on some of the technologies, components, demonstrators and pilots that have been or will be delivered by a set of relevant EU-funded research projects in the area of Internet of Things. The information presented summarizes the assessment carried out within the Activity Chain 3 (AC3) "Application scenarios, Pilots and Innovation" of the Internet of Things European Research Cluster (IERC). The AC3 is a mean for collective exploitation that Projects participating in the IERC Cluster can use; in fact, its main objective of is to assess the innovation and impact of IERC projects with the goal of fostering considerable commercial and industrial opportunities for European IoT. Even though it cannot be exhaustive of the overall European situation the following overview presents a very comprehensive assessment of the maturity of the tangible outcomes delivered by eleven EU-funded; namely, ALMANAC [2], ClouT [3], OSMOSE [4], RERUM [5, 6], SMARTIE [7], SocIoTal [8], VITAL [9], BUTLER [10], iCore [11], IoT.est [12] and OpenIoT [13] is presented. At the time of writing, most of these projects are still in progress.

This chapter is organized as follows: a brief overview of the selected Projects is presented in Section 4.2, including an overview of the application areas addressed by the considered IoT research projects and a consortium analysis by country and type of organization; then, in Section 4.3 a summary of the projects' pilots and demonstrators delivered by the projects is given.

4.2 IoT Projects

In this section, the group of IoT projects being considered is presented. The remainder of this section presents an overall analysis on the ecosystems represented by the Projects and the application areas they address, followed by an overview of the demonstrators and pilots they deliver. Project-by-project detailed information is then reported in the subsections that complete this section.

In Figure 4.1, a map of the Organizations participating in the eleven Projects considered is depicted. As it is expected, the large majority of participants are from Europe. However, a strong collaboration with Japan can be seen. This is due to the fact that the ClouT project is part of a EU – Japan collaboration and because Japanese partners are also present in two other projects; namely, iCore, IoT.est.

A more detailed view of the map, considering only the European countries is shown in Figure 4.2.

The same data presented in Figure 4.3 grouped by Continents is shown in Figure 4.4. In Figure 4.3 a more detailed view of the European participation is shown. Spain, Italy and Germany are the Countries with highest involvement with 16%, 16% and 13% respectively, followed by France (7%), Greece (6%) and U.K. (5%).

In order to achieve successful results EU-funded projects' consortia are built upon strong partnerships of leading European and extra-European Organizations, including industries, universities and research centers, as well as and Cities and Public Administrations.

OPENHEATMAP

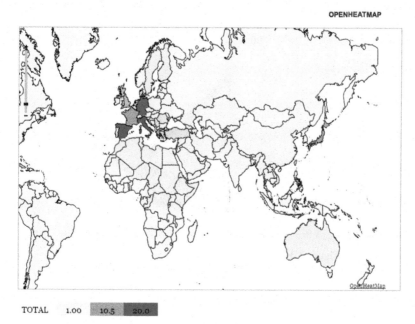

TOTAL   1.00   10.5   20.0

Figure 4.1   World countries participation in IoT EU-funded projects.

OPENHEATMAP

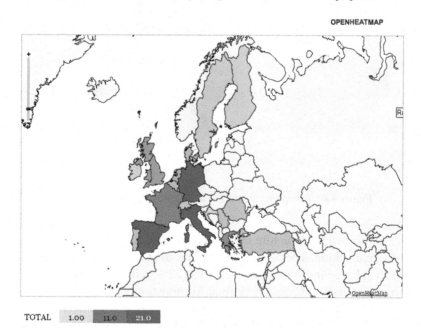

TOTAL   1.00   11.0   21.0

Figure 4.2   European countries participation in IoT EU-funded projects.

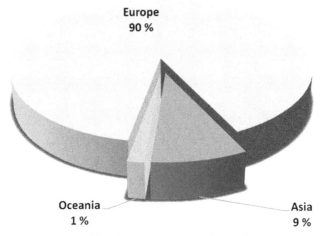

Figure 4.3 Consortium partners by continent.

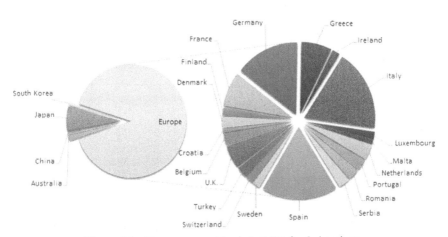

Figure 4.4 European countries in IoT EU-funded projects.

Figure 4.5 shows the structure of the eleven projects' consortia arranged by type of Organization. There is a rather balanced participation of Research Centers (25%), Large Industries (24%), SMEs (22%), and Universities (19%) with a minor presence of Cities and Public Administrations (10%).

IoT technologies are expected to foster innovation in a number of core European industrial sectors, including factory automation, sustainable energy,

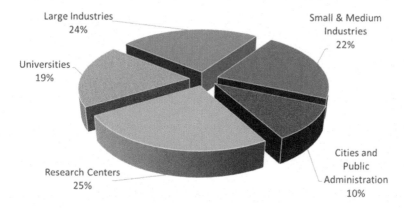

Figure 4.5 Consortium composition by Partner type.

automotive and aerospace, food production and distribution, environmental monitoring. Furthermore, the future IoT will be a cornerstone for the development of smart cities with more conscious citizens living in a more efficient and sustainable fashion.

A tentative preview of which of the sectors will be impacted most can be seen from the data presented in Figure 4.6, where the application areas addressed by the eleven research projects considered in this chapter are shown (for the sake of clarity, it is worth mentioning that the majority of the projects considered were funded under a specific "IoT for Smart City" FP7 call).

The remainder of this section details the relevant information on a project-by-project basis, enumerating the application areas targeted by each project together with a short description of the pilots and demonstrators deployed.

4.2.1 ALMANAC

The ALMANAC project will develop an IoT platform that promotes integrated smarter city processes for green, citizen-centric and sustainable urban ecosystems.

The open software ALMANAC Smart City Platform (SCP) enables seamless integration of devices, services, and private and public data, as well as federation of existing services. This is achieved by using a set of basic building blocks that ease third-party application development.

The SCP also enables interoperation of different communication networks and heterogeneous IoT technologies. Experimentation of selected Smart City services and applications will be carried in the city of Turin, Italy.

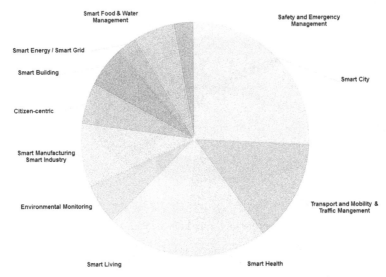

Figure 4.6 Application areas addressed by IoT research projects.

4.2.1.1 Application Areas
The applications areas covered by the projects are as following:

- Smart City
- Water Management
- Waste Management
- Citizen-Centric

4.2.1.2 Pilots and Demonstrators
4.2.1.2.1 *Smart Waste Collection Field Trial*
ALMANAC is currently developing a field trial in Turin, Italy, implementing an innovative waste collection system. The initial deployment foresees the installation of: i) fill-level sensors in 2 Underground Ecological Islands (UEI), ii) controlled access modules for a subset of the monitored waste bins in the selected UEIs, and iii) weight sensors on-board of the waste collection trucks.

The smart waste collection field trial integrates information from an issue reporting and management application developed by the project using the ALMANAC SCP, which enables citizens to report issues and irregularities in the waste management service. The same application provides a feature that, based on information from sensors deployed in the field and feedback from citizens, allows waste collection routes modification and can be used by the service operator to optimize the waste collection service.

An extension of the field trial has been already evaluated and will be possibly deployed to include additional UEIs and a number of street level waste bins in different locations in the city of Turin.

4.2.1.2.2 *Smart Water Capillary Network*
Proof of Concept demonstrator of the capillary network developed by ALMANAC for a water metering application. The Capillary Network provides the infrastructure to collect the data originated by different devices (sensors, meters, etc.) and ensures their collection in a ETSI M2M compliant service platform. The PoC consists of a smart water meter (flow meter and Ph sensor) sending data periodically using standard protocol – through a concentrator with IP connection – to the ALMANAC SCP, enabling both real-time and historical monitoring of water consumption data.

4.2.1.2.3 *Collaborative Citizen-centric application*
ALMANAC is developing a citizen-centric application with the goal to demonstrate the SCP capabilities to provide access to open public data and integrate third-party services relevant to the citizens with other services provided by the platform. In order to achieve this, ALMANAC is implementing a co-design approach in which a selected group of citizens have been engaged to participate in the definition of the application and will test the application prototype providing relevant feedback to further refine it and assess its acceptance.

4.2.2 ClouT

The overall concept of ClouT, a joint European-Japanese project, is leveraging the Cloud Computing as an enabler to bridge the Internet of Things with Internet of People via Internet of Services, to establish an efficient communication and collaboration platform exploiting all information sources to make the cities smarter and to help them facing the emerging challenges such as efficient energy management, economic growth and development.

4.2.2.1 Application Areas
- Smart City
- Safety and Emergency Management
- Smart Transport and Mobility
- Citizen-centric
- Smart Living

4.2.2.2 Pilots and Demonstrators
4.2.2.2.1 *Fujisawa field trial: Surfboard + Smile Coupon*

The purpose of "Eno-kama Info Surfboard" and "Smile Coupon" field trial applications is to provide event information and traffic information in Fujisawa area to tourists in real time, in order to create a new route for tourists for sightseeing. These two applications are based on ClouT architecture and are deployed in Kamakura station in cooperation with Enoshima Electric Railway Co., Ltd. While "Surfboard" is providing city context information, "Smile coupons" is providing discounts at the local shops. The more you smile the greater the discount applied.

4.2.2.2.2 *Mitaka field trial: Paw collection*

This application will help motivate elderly citizens to go out by providing interesting information such as event or city information provided by citizens, stores, event organizers and city. A Social Network Service (SNS) with integrated Sensor Data will motivate elderly people to go out more frequently and take longer walks, preventing them to need nursing care, by using an application called "Paw Collection". Paw is a kind of People's experience that is posted as an article with sensor data on the SNS.

4.2.2.2.3 *Genova field trial: "I don't risk" (SEM[1] application context)*

The purpose of this field trial is to inform citizens about good practices and general information about environmental risks and emergency situations, in order to reduce individual exposure to these risks. "I don't risk" application uses environmental and weather data from weather sensors, hydrometers, webcams, etc. and provides information to the citizens and the Civil Protection agency of Genova city.

4.2.2.2.4 *Santander field trial: Traffic Mobility Management*

This field trial will enable citizens and visitors of Santander City to get access to enhanced urban mobility experience and to leverage city transportation resources efficiently. Vast amount of information generated in the city is processed in order to generate real time alerts and events about relevant city information.

---

[1] Safety and Emergency Management.

4.2.2.2.5 *Sensorized garbage collection cars*

Environmental sensors (CO, O_3, NO_2, pollen, luminance, humidity, UV, temperature and ambient noise) installed in garbage collection cars. By having garbage collection cars sensorized, the project is able to provide citywide information of the atmosphere to citizens and visitors in Fujisawa city.

4.2.2.2.6 *Pace of the city*

Provide users with the capacity of utilizing their mobile phones to send in an anonymous way, physical sensing information, e.g. GPS coordinates, compass, environmental data such as noise, temperature. Users can also subscribe to services such as "the pace of the city", where they can get alerts for specific types of events currently occurring in the city. Users can themselves also report the occurrence of such events, which will subsequently be propagated to other users that are subscribed to the corresponding types of events.

4.2.3 OSMOSE

OSMOSE provides a roadmap and a technology platform that should support the transition and implementation of European SMEs of new business models and strategies in the digital world. The impact of OSMOSE is to provide already a middleware that is capable of addressing an increased asset connectivity to a digital enterprise. Through sensing enterprise capabilities OSMOSE should unlock new business models opportunities.

4.2.3.1 Application Areas

- Smart Manufacturing

4.2.3.2 Pilots and Demonstrators

4.2.3.2.1 *OSMOsis applications for Aerospace Domain*

Proof of concept referred to a product operations monitoring and control use case using the flight simulators in AgustaWestland Italy.

The goal of this PoC is to assure the training continuity and continuously improve the system reliability focusing on software snags faster assessment and resolution and hardware faults prevention.

4.2.3.2.2 *OSMOsis applications for Automotive Domain*

Engine Power Components (EPC) pilot in Spain is a proof of concept of OSMOSE. The proposed PoC is dedicated to manage the whole production process of camshafts, from its provisioning to its production, distribution and remanufacturing, if needed. The camshaft will be digitalized from its origin to its destination and all this information will be stored to keep the track of the whole life-cycle of a camshaft, reducing the risk of delivering camshaft with defects and having more data available to make decisions in real time about the changes in the process.

4.2.4 RERUM

RERUM increases the trustworthiness of IoT providing an overall security, privacy and trust framework to address the citizens requirements for advanced, reliable, resilient and secure smart city applications that respect their privacy improving both devices and middleware functionalities.

4.2.4.1 Application Areas
- Smart City
- Smart Transport and Mobility
- Smart Environmental monitoring
- Smart building/Smart Energy
- Deployments, Pilots and Demonstrators

4.2.4.2 Pilots and Demonstrators
4.2.4.2.1 *RERUM Mote (ReMote[2])*
ReMote is an innovative hardware IoT platform fully designed and developed based on the requirements set by the RERUM project. It is both powerful and low-power so that it can run the device-embedded security, privacy and reliability RERUM mechanisms, while consuming very low energy.

4.2.4.2.2 *Smart Transportation*
This pilot utilizes mobile devices to gather traffic information throughout city areas in a privacy-preserving way, without disclosing any type of personal information of the users from their mobile phones. For the pilot, the devices will be installed on buses that traverse around the city area to measure the traffic at specific roads. Volunteer citizens will also be able to participate in the pilot by downloading the RERUM application and installing it on their devices as they move around the city.

---

[2]http://zolertia.io/products

4.2.4.2.3 *Smart Environmental monitoring*

This pilot deals with the deployment of a secure and reliable system for gathering environmental monitoring measurements throughout city areas. This will be done in a secure and trustworthy way either by deploying sensor nodes at specific fixed locations or by installing sensors on top of buses and gathering the measurements at every bus stop. The goal is to ensure that no malicious users can intervene in the transmission of the measurements or gain unauthorized access to the system services.

4.2.4.2.4 *Home energy management*

This pilot deals with the deployment of sensors for securely measuring the energy consumption of specific household appliances (e.g. air conditioning systems, PCs, lights, etc.) and extracting information and usage patterns so that guidelines for minimizing the energy consumption will be provided. At Heraklion the system will be installed in two buildings one old and one new "green" building. At Tarragona the system will be installed at municipal offices. This pilot deals with both improving the security of the system and preserving the privacy of the user data, ensuring that no external party can identify when the user is at home or usage patterns for the devices.

4.2.4.2.5 *Comfort quality monitoring*

This pilot utilizes sensors for measuring securely the indoor air quality at buildings and extract information so that guidelines for improving the air quality will be provided. At Heraklion the system will be installed in the same two buildings as with the home energy management use case for comparing the results. At Tarragona the system will be installed at the municipal offices. Security of the system, privacy of the user data and trust in the system are key RERUM advances in such an application.

4.2.5 SMARTIE

The SMARTIE project works on security, privacy and trust for data exchange between IoT devices and consumers of their information. Results are demonstrated in smart cities in Germany, Serbia and Spain. The vision of SMARTIE is to create a distributed framework to share large volumes of heterogeneous information for the use in smart-city applications, enabling end-to-end security and trust in information delivery for decision-making purposes following data owner's privacy requirements.

4.2.5.1 Application Areas
- Smart City
- Smart Transport and Mobility

4.2.5.2 Pilots and Demonstrators
4.2.5.2.1 *Augmented Reality Based Smart Transport Service*

Improving the management of the public transportation network in the city of Novi Sad to promote and encourage the use of sustainable transport modes and to provide time and cost benefits to travelers. Focusing initially on 2 routes within a city public bus transport network operated by a local transport company JGSP. Bus stops will be equipped with the Augmented Reality (AR) markers in the form of an image (e.g. logo or QR code). Devices to measure air pollution in the busses, an e-ticketing system and a mobile app providing touristic information and event suggestion, are also expected to be integrated to the pilot.

4.2.6 SocIoTal

SocIoTal designs key enablers for a reliable secure and trusted IoT environment facilitating the creation of a socially aware citizen-centric Internet of Things. It takes a citizen-centric approach towards creation of large-scale IoT solutions of interest to the society. SocIoTal provides secure and trusted tools that increase user confidence in the IoT environment.

4.2.6.1 Application Areas
- Smart City
- Smart Living
- Citizen-centric services

4.2.6.2 Pilots and Demonstrators
4.2.6.2.1 *Santander and Novisad trials*

Enabling citizens and developers to develop new services using SocIoTal toolset. The initial set of trials is based on the output from co-creation workshops and the IoT meetups held in several cities. Examples are monitoring of lifts in the buildings, measuring happiness of a city as well as sharing data generated by citizen owned devices and navigating through the routes accessible to disabled people.

4.2.7 VITAL

In the VITAL project, the future of Smart Cities, the project is developing a novel virtualization layer for the next generation of integrated and technology independent smart city operating systems in Europe.

4.2.7.1 Application Areas
- Smart City
- Smart Transport and Mobility

4.2.7.2 Pilots and Demonstrators
4.2.7.2.1 *IoT-supported Urban Regeneration*
Hosted in London's first Business Improvement District (B.I.D), namely CTU, in the Camden Borough of London. CTU will act as an Urban Regeneration Living Lab in order to develop and integrate regeneration-related services based on VITAL virtualization approach.

4.2.7.2.2 *IoT-enabled Smart Traffic Management*
Development and validation in the city of Istanbul. Traffic management and analysis functionalities based on multi-source data sets. End-users include citizens and the city authorities

4.2.8 BUTLER (Completed)

BUTLER aimed to design and demonstrate the first prototype of a comprehensive, pervasive and effective Context-Aware information system, operating transparently and seamlessly across various scenarios towards a unified smart urban environment.

4.2.8.1 Application Areas
- Smart City
- Smart Health
- Smart Transport
- Smart Living

4.2.8.2 Pilots and Demonstrators
4.2.8.2.1 *SmartOffice Trial*
Deployment of IoT technologies based on the BUTLER platform in three of the offices of the project partners. The three trials shared common functional requirements (information sharing, office wellbeing), all three sites

participated in a common PoC of IoT information sharing: coffee consumption data shared between the offices

4.2.8.2.2 *SmartShopping Trial*
Processing in real time of the city status for creating alerts for the merchants, to inform about potential presence of customers that fit with business profile.

4.2.8.2.3 *SmartParking Trial*
Smart Parking Management System. A group of users tested during various days the SmartParking devices, the reservation system and the mobile app.

4.2.8.2.4 *SmartHealth Trial*
Involving IoT technologies at home for health monitoring with the aim of helping people to control certain diseases. TECNALIA has developed different devices and has integrated them in the BUTLER platform (e.g. fall detector, emotion detector, medication intake assistant, telecare reporting service, videoconference for medical and risk prevention service).

4.2.8.2.5 *SmartTransport Trial*
Enabling public transportation systems use without taking care of pricing or ticketing leveraging on IoT solutions (i.e. e-ticketing, save child group monitor and tags). Real field trial took place in collaboration with TU Dresden ITVS and Fraunhofer IVI at AutoTRAM Extra Grand in June to October 2014.

4.2.9 iCore
iCore addressed two key issues in the context of the Internet of Things: abstraction of the technological heterogeneity deriving from the vast amounts of objects, while enhancing reliability; and considering the views of different users/stakeholders (owners of objects & communication means) to ensure proper application provision, business integrity and, therefore, maximize exploitation opportunities. The iCore proposed solution is a cognitive framework reusable for various and diverse applications.

4.2.9.1 Application Areas
- Smart City
- Smart Transport and Mobility
- Smart Health
- Smart Living

- Smart Manufacturing
- Smart Building

4.2.9.2 Pilots and Demonstrators
4.2.9.2.1 *Pilot at Trento Hospital*
The pilot is setup at the department of neonatology of the central hospital of Trento, Italy. The pilot addresses tracking of portable medical equipment (in the range of few tens of items) inside the unit as well as from/to other relevant units, such as gynecology and emergency service. Information about usage and movement of devices will be used to produce predictive models about statistical usage and location of items to support both, end-users (doctors and nurses as well as hospital procuring and maintenance departments) as well as indoor location developers for reducing energy consumption of their tracking system.

4.2.9.2.2 *Smart Tour in the City*
This trial aimed to apply aspects of cognitive management for IoT self-management in the scope of a Smart tour in the city application, involving actual, diverse users. Part of the trial took take place in Athens and involved users visiting different sites around the city. Another major part of the trial concerned the exploitation of the SmartSantander infrastructure for conducting experiments for the large scale evaluation and validation of the integrated iCore architecture and concepts.

4.2.9.2.3 *Smart Urban security*
Urban Security and VIP protection Demonstrator. A surveillance system focused on VIP protection during a visit within a big and crowded exhibition area. Police Control and Command (C2) truck close to the area with dedicated surveillance applications is monitoring the VIP visit through a deployed wireless (video and chemical) sensors network also connected to exhibition area CCTV system and chemical sensors. When a dirty bomb explodes generating toxic cloud dispersal, VIP evacuation is triggered and managed up to a decided exit according to threats tracking (toxic cloud, crowds). iCore cognitive platform embedded in C2 truck manages in real time optimal selection of video streaming and use of available WSN bandwidth.

4.2.9.2.4 *Smart Home*
IoT self-management aspects through cognitive functionalities in the scope of a Smart Home. The prototype comprises, apart from software components for

the various functional components, Arduino platforms combined with various sensors and actuators such as temperature, humidity, luminosity, body pulse and motion detection sensors, accelerometer, lamp, fan/heating and buzzer. Software technologies used for the implementation include RESTful Web Services, JSON, RDF, SPARQL, Sesame API, etc.

4.2.9.2.5 *Task-based Smart IoT*

Testbed environment that recommends appropriate tasks as a composition of IoT devices and their services. It is implemented in two rooms: a seminar room and a resting place. The seminar room has various smart objects like projector, screen, light, robot vacuum cleaner, flower pot, temperature/humidity/light sensor, and air conditioner (total 7 objects). The resting place includes smart board screen, smart fridge, and temperature/humidity/light sensor (total 3 objects). Basically the Task-based Smart IoT Prototype can interact with any user who has a smartphone installed with a client application. Based on scenarios including group tasks such as meeting, watching a movie, gaming, etc., the task-based Smart IoT Prototype properly supports up to 10 users.

4.2.9.2.6 *Smart City: Transportation*

Demonstrates the virtualization and use of ICT objects in Automotive industry, to create, configure and use mobility functions and services while driving and, in a seamless way, also in pre-trip and post-trip services, linking to smart home and smart meeting. Although the focus is on a single driver, data provision from several cars will also be addressed, for the mobility management in a smart city. Major aspects and challenges are the availability of objects within the vehicle and from the outside world, considering the vehicle as a complex and autonomous eco-system and not an always connected environment. Another topic addressed is context awareness using cognitive technologies.

4.2.9.2.7 *Smart Office*

Demonstrates the capability of managing the whole lifecycle of a meeting, from its organization to its execution and wrap-up, while re-using already existing IoT devices (smartphones, tablets, smart panels). This is achieved through the development of appropriate Virtual Object (VO) containers for these devices that enable them to become part of an iCore ecosystem, while appropriate Composite VOs at the back-end are able to monitor the meeting and provide the required functionalities for supporting a variety of service requests, ranging from sending out the meeting invitations to supporting the indoor navigation of participants to the meeting venue, the recording of the meeting and the "attention-span" management (smart break) to the eventual wrap-up of the meeting with the uploading of the meeting recordings to a designated area and the notification of the participants for their offline availability.

4.2.9.2.8 *Smart cold chain logistics*

The domain implies high complexity and high risks because food and pharmaceutical goods are exposed to increasingly long and complex supply chains with many dangers of poor temperature control, delays and physical mishandling. The prototype improves the transportation process by monitoring the state of the products during transportation and by early warnings when the goods are not stored according to clients' requirements.

4.2.10 IoT.est (Completed)

The IoT.est project demonstrates the whole service creation life cycle for IoT services. IoT services can be regarded as being, in principle, similar to any classical service with the marked distinction that part of the service instantiations relies on information obtained from IoT devices (sensor or other sources as well as actuation). Generation of test cases, testing, service-redefinition as well as monitoring will be automated (or at least semi-automated). The four phases of the IoT service life cycle are therefore enhanced with testing and monitoring facilities at different stages of the cycle.

4.2.10.1 Application Areas
- IoT service creation, testing and deployment

4.2.10.2 Pilots and Demonstrators
4.2.10.2.1 *IoT Services Testing*
The "EEBuildingSim" exposes four types of simulated IoT resources: temperature sensors, window actuators, AC units, as well as heaters, currently exposed as single atomic services. These services will be accompanied by a set of "smarter" atomic services which will include some form of embedded logic combining sensing and actuation into a single service: basic temperature regulator, advanced temperature regulator and follow the fire service.
Three scenarios considered for further PoC demonstrators:

- Emergency – Smart Events Scenario
- Energy – Energy Efficient Buildings Scenario
- Healthcare – Well Being Scenario

4.2.11 OpenIoT

OpenIoT is a open source platform supporting semantic interoperability between sensor data silos and also for enabling Internet of Things semantically-annotated services in the cloud.

4.2.11.1 Application Areas
- Smart City
- Smart Transport and Mobility
- Smart Health
- Smart Living
- Smart Manufacturing
- Smart Building

4.2.11.2 Pilots and Demonstrators
4.2.11.2.1 *IoT-Smart City – Crowdsensing Quality of Air Monitoring Trial*
Realized through an opportunistic Mobile Crowdsensing application involving volunteers carrying smartphones and air quality sensors that contribute the sensed data to the OpenIoT platform.

4.2.11.2.2 *IoT-Intelligent Manufacturing – Smart Industry Trial*
Means for dynamically selecting production process monitoring sensor information, as well as for structuring this information on KPIs and make them available in form of customized and created on the fly. Validated in the paper/packaging industry in processes like printing, die-cutting and gluing/folding.

4.2.11.2.3 *IoT Enabled (Smart) Campus Guide*
The University Smart Campus (synonymous with CampusGuide) is an application framework to support students, teachers and guest of a university. It offers features like information's about buildings and rooms, reservations of meetings rooms and workplaces, and collaboration between people.

4.2.11.2.4 *Silver Angel – IoT Enabled Living and Communication in Smart Cities*
The purpose of the Silver Angel application is to help ageing citizens live independently in their own homes, and to facilitate meeting more often with friends and relatives. It offers three Silver Angel services namely Smart Meeting, Issue Reporting and Alarms.

4.2.11.2.5 *IoT-Large Scale Deployments – Phenonet Trial*

'Phenonet' describes the network of wireless sensor nodes collecting information over a field of experimental crops. The term "Phenomics" describes the study of how the genetic makeup of an organism determines its appearance, function, growth and performance. Plant phenomics is a cross-disciplinary approach, studying the connection from cell to leaf to whole plant and from crop to canopy.

4.2.11.2.6 *Openlot middleware platform and Virtual Development Kit*

The OpenIoT middleware was released in May 2013. and from September 2013 was made fully available to the Open Source community for creating real time IoT services on demand and enable interoperability between vertical IoT solutions and interconnect data silos. The OpenIoT framework (BETA version v0.1.1) has been released via the github project management portal https://github.com/OpenIotOrg/openiot. Likewise the first version of the Virtual Development Kit – OpenIoT-VDK (running Linux) with the complete OpenIoT platform pre-installed and preconfigured for use/development has been released. It can be downloaded from the OpenIoT github wiki: https://github.com/OpenIotOrg/openiot/wiki/Downloads. The Virtual Development Kit (OpenIoT-VDK) developed and implemented, features the OpenIoT latest release i.e. v0.1.1 and it's size is 5, 7 GB. The OpenIoT-VDK facilitates the learning and use of the OpenIoT framework for an easy adoption and it is industry friendly under LGPL licence and totally open for academic purposes. The OpenIoT-VDK instance deploys the IoT service delivery model facilitating the validation of use cases by using the OpenIoT platform.

4.3 IoT Projects' Pilots and Demonstrators

In order to analyze the tangible outcomes of the group of IoT projects being considered, two subsets of outcomes will be presented demonstrators and pilots.

In this document, a demonstrator is intended as a system that attempts to emulate the full system or sub-system behavior, in order to test its main capabilities. A pilot instead, is understood as the deployment of the full-system, tested against a subset of the general intended end-users with the goal to better appreciate how the system will be used in the field in order to refine it, if necessary.

In this section, information regarding the demonstrators and pilots deployed by the group of IoT projects being following information is given:

- Name of the deployment and project
- Status of the deployment (P: planned, O: ongoing, F: finished)
- Location (country)
- Dimension (size of the deployment in number of users/devices/data set/area)
- Notes including additional information regarding the dimension and devices deployed by the project

In Table 4.1 and Table 4.2, the demonstrators and pilots from the selected group of IoT projects are presented respectively.

Table 4.1   Description of the Demonstrators of the considered IoT projects

Project	Demonstrator	Location	Notes
ALMANAC	Smart Water Capillary Network	Italy	Telecom Italia Labs
	Collaborative Citizen-centric application	Italy	Co-designed and tested by the SHARING[3] community
RERUM	ReMote	n.a.	Off-the-shelf IoT hardware platform
OSMOSE	Aviation PoC (AgustaWestland)	Italy	
	Automotive proof-of-concept scenario (EPC)	Spain	
iCore	Smart Home	Greece, Germany	
	Task-based Smart IoT	n.a.	seminar room
	Smart City: Transportation	Italy	
	Smart Office	UK, Spain	
	Smart cold chain logistics	n.a.	no specific location
IoT.est	IoT Services Testing	virtual machines	no specific location
OpenIoT	OpenIot middleware platform and Virtual Development Kit	n.a.	https://github.com/ OpenIotOrg/openiot

---

[3] http://www.sharing.to.it/sharingHtml.html

Table 4.2   Description of the Pilots of the considered IoT projects

Project	Pilot	Status	Location	Dimension	Notes
ALMANAC	Smart Waste Collection Field Trial	O	Turin, Italy	8	8 waste bins equipped with fill-level sensors and controlled access[4] modules, participants reporting issues using the mobile app
ClouT	Fujisawa field trial: Surfboard & Smile Coupon	O	Fujisawa, Japan	150	150 users plus sensors, SNS and websites
	Mitaka field trial: Paw collection	F	Mitaka, Japan	60	Users using the mobile app
	Genova field trial: "I don't risk"	F	Genoa, Italy	3000	Users from the Genoa City
	Santander field trial: Traffic Mobility Management	O	Santander, Spain	>1000	Traffic, environmental, parking and noise sensors
	Sensorise garbage collection cars	O	Fujisawa, Japan	>10	Sensors and GPS systems installed in cars
	Pace of the city	F	Santander, Spain	>100	Users in the city
RERUM	Smart transportation	P	Greece, Spain	40 + 30	40 mobile devices in buses, 30 users devices
	Environmental monitoring	O + P	Greece, Spain	50 + 20	Sensor nodes for environmental monitoring
	Home energy management	P	Greece, Spain	10 + 10	Sensors for energy consumption
	Comfort quality monitoring	P	Greece, Spain	10 + 10	Sensor nodes for environmental monitoring
SMARTIE	Augmented Reality Based Smart Transport Service	n.a	Serbia	15	Devices to measure air pollution, QR codes, mobile app
SocIoTal	Santander and Novisad trials	O	Spain, Serbia	>100	Participants, data collected from their devices and behaviors

(Continued)

[4]Only a subset of the waste bins will be initially equipped with controlled access modules.

Table 4.2 Continued

Project	Pilot	Status	Location	Dimension	Notes
VITAL	IoT-supported Urban Regeneration	P	UK	n.a.	Mobile Phones, FootFall Systems, CCTV cameras, various open data sources
	IoT-enabled Smart Traffic Management	P	Turkey	500 + 500 + 250 + 35	500 traffic cameras, 500 road sensors; 250 Bluetooth sensors; 35 weather sensors
BUTLER	SmartOffice Trial	F	France, Italy, Switzerland	25 + 15 + 25	Users in the three different locations
	SmartShopping Trial	F	Spain	100 + 150 + 100 + 400	100 restaurants, bars and hotels, 150 shops
	SmartParking Trial	F	Spain	40 + 20	100 merchants, 400 End-users
	SmartHealth Trial	F	Spain	50 + 1	4 parking lots, with 10 parking spaces per lot under control, 3 trials of 20 people
	SmartTransport Trial	F	Germany	50	50 final users, 1 Doctor
iCore	Pilot at Trento Hospital	F	Italy	>50	50 tags have been deployed on as many portable devices inside the hospital
	Smart Tour in the City	F	Greece	75	75 traffic and environment sensors
	Smart Urban security	F	France	25 + 25 + 4 + 10	Simulated exhibition area CCTV system (25 cameras); Simulated exhibition area chemical sensors network (25 sensors), simulated portable video cameras and chemical sensors (4 undercover policemen); WSN with 10 nodes.

OpenIoT	IoT-Smart City – Crowdsensing Quality of Air Monitoring Trial	O	Croatia	16000	Data set of 16.835 measurements, crossed in distance of 758.6km, covered area 144 km^2
	IoT-Intelligent Manufacturing – Smart Industry Trial	O	Greece	20	20+ sensors in manufacturing plants (optical diffusion sensors, image sensors, laser barcode scanners temperature sensors, weight sensors and RFID sensors)
	IoT Enabled (Smart) Campus Guide	O	Germany	20	20+ Mobile Phones, RFID tagged objects, data sources/streams
	Silver Angel – IoT Enabled Living and Communication in Smart Cities	O	Germany	n.a.	Mobile phones (for citizen reporting) used along with open data from public administration
	IoT-Large Scale Deployments – Phenonet Trial	F	Australia	n.a.	–

4.4 Summary

A group of IoT research projects were presented in this chapter. An overall as well as a detailed analysis of the target application areas they address, their consortium structure (i.e. by country and type of organization) and the pilots, demonstrators and other tangible outcomes they intend to deliver by the date of completion was presented.

4.5 List of Contributors

Project	Name	Organisation
ALMANAC	Claudio Pastrone	ISMB
ClouT	Levent Gurgen	CEA-LETI
	Jose Antonio Galache	Universidad de Cantabria
OSMOSE	Sergio Gusmeroli	TXT e-solutions
	Roberta Caso	Reply
	Gabriella Monteleone	Piksel
RERUM	Elias Tragos	ICS-FORTH
SMARTIE	Boris Pokric	DunavNet
SocIoTal	Klaus Moessner	Surrey University
	Srdjan Krco	DunavNet
VITAL	John Soldatos	Athens Information Technology
	Gregor Schiele	INSIGHT Centre
	Martin Serrano	INSIGHT Centre
BUTLER	Bertrand Copigneaux	Inno Group
iCore	Raffaele Giaffreda	CREATE-NET
	Vera Stavroulaki	UPRC
	Panagiotis Vlacheas	UPRC
	Dimitris Kelaidonis	UPRC
	Vasilis Foteinos	UPRC
	Panagiotis Demestichas	UPRC
	Stylianos Georgoulas	UNIS
	Klaus Moessner	UNIS
	Massimo Barozzi	TRILOGIS
	Giuseppe Conti	TRILOGIS
	Nicola Dorigatti	TRILOGIS
	Stefano Piffer	TRILOGIS
	Byoungoh Kim	KAIST
	Stephane Menoret	THALES
	Andrea Parodi	M3S
	Michele Stecca	M3S
	Michele Provera	CRF

	Septimiu Nechifor	SIEMENS
	Dan Puiu	SIEMENS
IoT.est	Thomas Gilbert	Alexandra Instituttet
OpenIot	Danh Le Phuoc	INSIGHT Centre
	Salma Abdulaziz	INSIGHT Centre
	John Soldatos	Athens Information Technology

Bibliography

[1] O. Vermesan and P. Friess, "Internet of Things Applications – From Research and Innovation to Market Deployment," River Publishers, 2014. [Online]. Available: http://www.internet-of-things-research.eu/pdf/IoT-From%20Research%20and%20Innovation%20to%20Market%20Deployment_IERC_Cluster_eBook_978-87-93102-95-8_P.pdf

[2] ALMANAC EU Project, "ALMANAC Project Website," [Online]. Available: http://www.almanac-project.eu/news.php

[3] ClouT EU Project, "ClouT project website," [Online]. Available: http://clout-project.eu/. [Accessed 05 2015].

[4] OSMOSE EU Project, "OSMOSE Project Website," [Online]. Available: http://www.osmose-project.eu/. [Accessed 05 2015].

[5] RERUM EU Project, "RERUM Project Website," [Online]. Available: https://ict-rerum.eu/. [Accessed 05 2015].

[6] E. Z. Tragos, A. Angelakis, A. Fragkiadakis, D. Gundlegard, C.-S. Nechifor, G. Oikonomou, H. C. Pohls and A. Gavras, "Enabling reliable and secure IoT-based smart city applications," in *Pervasive Computing and Communications Workshops (PERCOM Workshops), 2014 IEEE International Conference on*, 24–28 March 2014.

[7] SMARTIE EU Project, "SMARTIE Project Website," [Online]. Available: http://www.smartie-project.eu/

[8] SocIoTal EU Project, "SocIoTal Project Website," [Online]. Available: http://sociotal.eu/. [Accessed 05 2015].

[9] VITAL EU Project, "VITAL Project Website," [Online]. Available: http://vital-iot.eu/. [Accessed 05 2015].

[10] BUTLER EU Project, "BUTLER Project Website," [Online]. Available: http://www.iot-butler.eu/. [Accessed 05 2015].

[11] iCore EU Project, "iCore Project Website," [Online]. Available: http://www.iot-icore.eu/. [Accessed 05 2015].

[12] IoT.est EU Project, "IoT.est Project Website," [Online]. Available: http://ict-iotest.eu/iotest/. [Accessed 05 2015].

[13] OpenIoT EU Project, "OpenIoT Project Website," [Online]. Available: http://cordis.europa.eu/project/rcn/101534_en.html. [Accessed 05 2015].

5

Industrial Internet of Things and the Innovation Processes in Smart Manufacturing

Sergio Gusmeroli[1], Oscar Lazaro[2], Željko Pazin[3], Chris Decubber[3], Paolo Pedrazzoli[4], Didier Vanden Abeele[5], Sara Tucci-Piergiovanni[5], Jacopo Cassina[6], Daniele Cerri[1], Sergio Terzi[1], June Sola[2], Ignacio Arconada[7], Klaus Fischer[8], Artur Felic[9], Michele Sesana[10], Roberto Sanguini[11], Lina Huertas[12], Domenico Rotondi[13] and Joe Cecil[14]

[1]Politecnico di Milano, Italy
[2]Asociacion De Empresas Tecnologicas Innovalia, Spain
[3]European Factories of the Future Research Association EFFRA, Belgium
[4]Scuola Universitaria Professionale della Svizzera Italiana SUPSI, Switzerland
[5]CEA List, France
[6]Holonix s.r.l., Italy
[7]TRW Automotive, Spain
[8]German Research Center For Artificial Intelligence – DFKI, Germany
[9]CAS Software AG, Germany
[10]TXT e-Solutions SpA, Italy
[11]AgustaWestland SpA, Italy
[12]The Manufacturing Technology Centre MTC, UK
[13]FINCONS SpA, Italy
[14]Oklahoma State University, US

5.1 IIoT for Manufacturing: Key Enabler for 4th Industrial Revolution

Manufacturing industry is the driving force for EU renaissance and recovery from the economic crisis. This chapter aims at investigating the enormous innovation potential of IoT technologies when fully adopted not just in the production of physical goods, but in all activities performed by Manufacturing Industries, both in the pre-production (ideation, design, prototyping, 3D printing) and in the post-production (sales, training, maintenance, recycling) phases (the SMILE challenge).

The first sub-chapter includes the Factories of the Future PPP perspective (EFFRA, European Factories of the Future Research Association) to IoT for Manufacturing.

The following four sub-chapters will proceed with short descriptions of four IoT for manufacturing FP7 research projects, each addressing different phases of the Product Lifecycle (design, manufacturing, operations, and maintenance) and the impact of IoT to their industrial cases.

Moreover, the importance to build and sustain a multi-disciplinary business ecosystem for IoT adoption in Manufacturing is described as well as the US perspective to IoT for Manufacturing provided by the IGNITE initiative.

Finally, a conclusive sub-chapter will report a synthesis of a recent EC-funded study identifying research-innovation-market challenges for IoT/Cloud combination in manufacturing. The resulting synthetic view of the study confirms that Smart Manufacturing is the most promising and disruptive sector for IoT adoption in the next few years, in combination with other KETs for manufacturing, such as 3D Printing and Cyber Physical Production Systems.

5.2 IoT in the Factories of the Future PPP and Digital Manufacturing: The EFFRA Perspective

The convergence of cloud and IoT technologies will facilitate the development of factories of the future and the realisation of digital manufacturing. These future manufacturing plants will comprise numerous devices, physical and virtual smart objects, internally and externally interconnected, by dynamically enabling configuration and monitoring of the operational capabilities of the plant, or networks of plants, quality control and efficiency improvement. Additionally, the traditional, fragmented processes of design, production and customer fulfilment will be replaced by a close-loop management of the end-to-end design-to-customer fulfil, where cycles are shorter and products are designed based on customer requirements (customer-focused manufacturing). Here, the processes do not finish with product delivery; the product-service provides information for the maintenance services and for continuous design of products and processes. The sensors in machinery and manufacturing services developed will facilitate the operational performance model for predictive maintenance of the machinery.

A global plant floor requires that the network of production facilities operates as a single virtual plant. Operations require individual plants' centralization control capabilities based on real-time information, multi-plant manufacturing execution systems (MMES) and major integration and visibility on supply customer ecosystems based on enterprise manufacturing intelligence (EMI) platforms. Additionally, increased control and supervision requires the improvement and acceleration of decision-making capabilities based on real-time information, interoperability between systems and collaborative decision making. This environment requires adaptive and scalable architectures to support real-time data for operational management, supply-chain execution and collaborative decision-making. Scalable and multi-enterprise architectures are needed for: managing the operations of networks of organizations in the same supply chain; connecting MES and business processes in real time; establishing new business models based on secure cloud services.

As the IoT expands, cybersecurity will have to be considered at every point, and common, sector-specific threats will need to be identified. Security requirements that are unique to CPS will have to be determined. A risk management framework and methodology to enable, assess, and assure cybersecurity for adaptive and smart manufacturing systems will have to be established; adaptable computational and storage tools including methods for protection and security of intellectual property will have to be identified, developed, and deployed. Novel information security concepts and/or approaches, such as turning properly constructed interfaces from attack surfaces into cyber-defence surfaces, offering explicit and implicit design guarantees, and providing security as a class of interface guarantee, will have to be explored. It will be much more effective and ultimately cheaper to secure smart manufacturing systems at the engineering design phase, rather than later. The economic and technical viability of possible integration with legacy systems as well as existing open source applications and tools will also have to be assessed.

5.2.1 IoT & Cyber-Physical Production Systems

As illustrated in the previous section, the factories of the future is being subject to a profound transformation. Such transformation will not be limited to the physical world and the manufacturing line. A digital revolution will also take place in the digital domain. Factories will witness the prevalence of Internet technologies also at manufacturing level; the mass deployment of Cyber-Physical Systems for monitoring & controlling will see the use of the Big Data capabilities everywhere (world, enterprise, shop-floor). The role of ICT will be instrumental addressing the increased complexity that manufacturing industries will have to face at many levels; e.g. increased product customization, largely dynamic delivery requirements, agile and rich collaboration patterns and networks of different technical disciplines and organizations. These requirements coupled with the advent of innovative manufacturing technologies such as additive manufacturing or high precision zero-defect manufacturing call for robust interoperability solutions that integrate the factory with the environment, i.e. the urban context, data integration standards amenable to IoT and linked tool chains that move away from vendor-lock-in, monolithic systems and envision mobile & cloud native support. The development of such smart manufacturing environment is highly dependent on the development and integration of IoT capabilities in an industrial context. Recalling a recent road-mapping effort; i.e. Pathfinder, IoT-based manufacturing must abandon the current classical approach to industrial automation – see figure below.

The current approach is deemed by industrial key players and RTD experts to be inadequate to cope with current manufacturing trends and needs to consequently evolve. The intrinsic existence of smart interconnected devices defies the concept of rigid hierarchical levels, being each one of these devices capable of complex functions across all layers. Thus it should be introduced an updated version of the pyramid representation, where the field level features

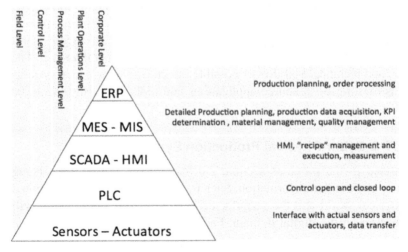

Figure 5.1   Traditional automation pyramid.

smart-objects capable of articulated functions (thus in contact with all the pyramid layers) while still the hierarchical structure is preserved. This leads towards the Industrial IoT automation pyramid that will be the basis for future Cyber-physical production systems (CPPS).

5.2.2 CPPS Architectures Design Drivers for Scalable, Adaptive and Smart Manufacturing Systems

Future ICT tools and technologies will give companies multiple opportunities, such as increases in efficiency and quality throughout value chains, the exploitation of additional markets, and manufacturing that is highly responsive to changing market and customer demands. Smart manufacturing will exploit advances in wireless sensor technologies, machine-to-machine (M2M) communication and ubiquitous computing, that would allow track-and-trace and monitor each individual stage of the production.

Thus, CPS will provide a shared situational awareness to support network-centric production by closing the loop between the virtual world and the physical world. In order to exploit the full potential of CPS, various existing ICT systems have to be integrated, adapted to the industrial needs, and deployed on the shop floor:

The evolution and design of ICT architectures to meet the demands of future CPPS capable of delivering the manufacturing competitive advantages described above will be driven by the **FISAR** design principles (**F**lexible system-component integration, **I**nteroperable among systems and

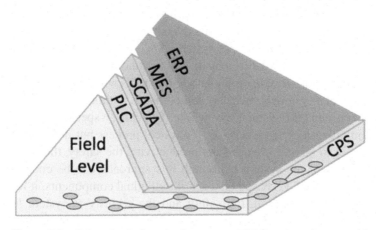

Figure 5.2 Cyber-physical production system (CPPS) automation pyramid.

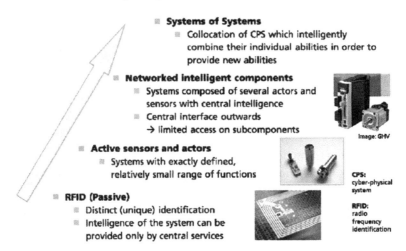

Figure 5.3 Stages of development in CPS capability.

components, **S**calable, **A**daptable to varying governance structures, **R**eal-time capabilities).

The design and development of CPPS is intrinsically complex with characteristics unique to the deployment of IoT technologies in an industrial context. The advent of the CPPS requires developing the material and software components, middleware, operating systems beyond the existing technologies. Because these systems interact directly with the physical environment their hardware and software must be reliable, reconfigurable, and, for the most critical, be certified, from components to the whole system. These complex systems have to present one degree of reliability/confidence, which is lacking to most of the current infrastructures. The oversizing is at present the way the most used for the certification of safe system.

Research is nowadays organised by disciplines and the systems are developed with a large set of formalisms and modelling tools without links between them. Each part of a system highlights characteristics without considering the other components and the systems as a whole. Typically, a specific formalism will represent either a cyber-process or a physical process but not both. The expertise is split to the detriment of the productivity, the safety, the security and the efficiency of the system. Although this approach can be enough to support a vision of the CPS based on a set of individual components, it raises a problem for the verification, the safety and the security at system level as for the interactions between components. To allow a design and a fast deployment of the CPS, it is necessary to develop innovative approaches to define architectures, which make possible a transparent integration of the elements of control, communication and processing. Making manufacturing smart implies that on one hand all devices in factory have to become smart, or smarter, and furthermore that they cooperate in order to provide smart functionalities. But these new functionalities shall not come at the price of decreasing plant safety, and in a connected world, safety necessarily imply cyber-security. The biggest challenge is that flexibility, safety and security requirements are usually competing with each other and refraining many smart manufacturing ideas from becoming a reality. So, the challenges for the engineering of CPPS are significant and range from the actual complexity of modular systems of systems engineering but also deal with the performance and reliability of SOA-EDA architectures, which increasingly demand a better integration between IoT and cloud domain. Moreover, cyber-security and in some cases mix-criticality of the system impose additional constraints that cannot be left aside.

5.3 Product Design and Engineering in the IoT Era: The LINKEDDESIGN Project

Over recent years, the context where companies operate has dramatically changed, forcing companies to revise their business models and to introduce new paradigms like sustainability. Sustainability is now considered a strategic must-have, because it enables profitability in the long run and a competitive differentiation [1]. Furthermore, companies now realize that sustainability improvements can have an immediate impact on the bottom line, not to mention propelling growth by spurring innovation and the creation of new products and services [2]. One way to pursue sustainability is to take a product life cycle approach, enabling a more efficient and effective use of limited financial and natural resources [3].

A life cycle approach enables product designers, service providers, government agents and individuals to make choices for the longer term, and it avoids shifting problems from one life cycle stage to another. Furthermore, experts from industry, government and other organizations agree that making life cycle approach part of the way, de-signing products, developing services, making policies will help to reverse some world damaging trends [4].

Companies are under pressure to create sustainable products, not only from consumers but also from governments, retailers and suppliers. As natural resources become scarce, companies must consider the long-term sustainability of their business models and broaden their approach to consider their total impact on the environment. Therefore, product life cycle approach can be a powerful growth engine [5].

To summarize, product life cycle approach enables to pursue sustainability, considering the whole product life cycle and not only what companies/individuals realize in a single stage. Furthermore, sustainability is now considered by companies not an imposition, but a competitive and long-run profitable leverage. Different companies, mainly in the automotive sector, like for example Volkswagen [6], BMW [7] and FCA Group [8], are pushing sustainability. In this context, focal companies of supply chains, as the previously cited ones, are driving their suppliers, employees, retailers and users towards the sustainability, in order to reduce the impact along the whole product life cycle.

Focus of this chapter is to consider and analyse manufacturing systems and their suppliers. Production phase is one of the most critical phases, concerning sustainability. Indeed, companies that use both environmentally and economically sound manufacturing practices can gain significant competitive advantages [9]. Furthermore, the chapter mainly considers the design process, due to its high influence on costs and environmental impacts generated along the whole manufacturing systems life cycle. Different empirical studies [10–13] state that product design represents 5–10% of life cycle cost; however, product design influences up to 80% of life cycle cost. Rebitzer et al. [14], instead, reports the same consideration about the environmental impacts.

In order to evaluate costs and environmental impacts generated during the whole life cycle, it is possible to use Life Cycle Costing (LCC) and Life Cycle Assessment (LCA) methodologies. Life Cycle Costing is described as the methodology that enables to evaluate the total cost of ownership of capital equipment, including its cost of acquisition, operation, maintenance, conversion and/or decommission [15]. Life Cycle Assessment, instead, is a methodology to assess environmental impacts associated with all the stages of a product's life from-cradle-to-grave [16], described in the standard ISO 14040 [17].

LCC and LCA methodologies are very good in comparison and estimation of few products or alternatives; however, when the number of alternatives increases, they are not able to support the decisions and the decision makers in a good way. In the case of manufacturing systems, composed by hundreds of different products or components, each with some alternatives, LCC and LCA are not able to support decision makers in the choice of the optimal system configuration, which minimize life cycle costs and environmental impacts.

Furthermore, LCC and LCA methodologies need to have real data from the system life cycle, in order to be more precise and accurate. Indeed, estimated or statistical data enable to reduce the time needed to perform a LCA study, losing accuracy.

Creating a tool able to collect data from the field and able to perform a LCA study with the same time, but a better accuracy, could be an interesting challenge to be faced. The aim of the chapter is to cover the gaps previously described, proposing a closed loop framework, completed by tools, for the improvement of industrial systems' sustainability.

Next paragraph describes the generic life cycle of a system and the closed loop framework, explaining how to apply it and giving a brief overview of different tools. Last section, instead, concludes the chapter.

5.3.1 IoT-Enabled Closed Loop Framework

In this section, the life cycle of a generic system life cycle is described, referring to Gera Model [18]. It identifies the following stages: identification, concept, requirements, preliminary design, detailed design, implementation, operation, possible redesign activities, and decommission.

Until requirements phase, customer and supplier must work together in order to establish system requirements. Then, supplier prepares a preliminary design of the industrial system, and usually this phase concludes with a proposal to the customer. Customer evaluates proposals of different suppliers; key drivers are the life cycle costs (and life cycle environmental impacts), and the best proposal gets the order. This is the most critical life cycle phase. If the order is won, the life cycle continues with the detailed design of the industrial system. Implementation phase represents the manufacturing and assembly of the industrial system at the customer plant and this phase concludes with the ramp up of the system. During Operation phase the system is fully operating. Furthermore, it is possible that supplier re-designs industrial system, in order to hit new customer needs. Finally, during Decommission, system's conditions are evaluated, in order to decide which is the best option (reuse as is, conversion to a new state, dismissal, etc.). The most critical phases are preliminary design and operation; indeed, it is important during the preliminary design to configure the optimal life cycle oriented solution, in order to get the customer order. Furthermore, it is important to collect information, returning valuable knowledge to keep under control the existing system and to improve the design of next systems. For these reasons, Gera Model is chosen, because it is more focused on design and operation rather than the other life cycle ones. The project has developed a reference framework for IoT-enabled closed loop design, completed by tools, along the whole life cycle of an industrial system.

The first tool is about the configuration of the suitable industrial system, in order to hit the customer's needs. In detail, the tool has to find the configuration that minimizes life cycle costs and environmental impacts and respects the customer's technical requirements.

The tool is based on NSGA-2 (Non dominated Sorting Genetic Algorithm) and implemented in a Java framework, in order to create a user interface and a back end engine. Back end engine is developed on JMetal [19], a library containing meta-heuristic models, like NSGA-2, to solve multi-objective problems. User interface enables the definition of the objective functions (which objectives to con-sider, like LCC, or LCA, or LCC and LCA) and of the technical requirements. Furthermore, it is possible to define how many alternatives are possible for each component of the system, within the Design Space Definition.

The tool checks and validates data inserted by user, in order to avoid errors. Finally, the algorithm can run. Tool displays algorithm results in an output window, which reports: (i) the value of the objective functions and (ii) all the optimal configurations (which components are selected to realize the system).

The second tool is about the collection of data from the field, during the operation phase. It is possible to collect tons of data and information from the

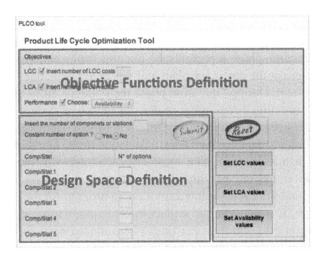

Figure 5.4   Configuration Tool User Interface.

different sensors installed on the system's components, machines or stations. Furthermore, the tool is also able to collect information by the functional subgroups of the machine/station, enabling to understand which ones influence the overall performance.

The issue is how to return all the extracted information in a valuable and useful way for the decision makers. The tool is composed by user interface, where the decision makers can interrogate the database, in order to receive the desired information (output), and by back end logic, in order to answer to the interrogations from the user interface. The back end recovers data and information from the PLC databases, which collect information by different system's sensors.

The main benefits of this tool are: (i) the real time monitoring of the system, keeping costs and environmental impacts of the existing system under control, and (ii) valuable knowledge for the design improvement of the next systems, in terms of performance and sustainability dimensions.

The tools return valuable knowledge, useful for the design of the next system. It enables system improvements, in terms of reduction of costs and environmental impacts, understanding which are the most critical components/stations/machines.

The tool is able to summarize the main system's parameters in order to verify quickly the performances of the system. It is possible to set different indicators, according to the customer needs, like the average cycle time, the mean time between failures (MTBF) and the mean time to repair (MTTR), the availability and the overall equipment effectiveness (OEE).

Decision makers can visualize detailed information, visualizing different pages that recap the main performances in a numerical and graphical way. For example, the page about availability shows the following parameters: availability, mean time between two failures (MTBF) and the mean time to repair the problem (MTTR). Furthermore, it shows the machine states.

5.3.2 Discussion

The chapter aims to propose an IoT-enabled closed loop framework for the improvement of industrial systems' sustainability, in terms of life cycle costs and environmental impacts. Introduction section describes the current context, where sustainability is now considered a must have and competitive leverage. Life cycle approach has been identified to pursue sustainability, considering the whole product life cycle and not only what companies/individuals realize in a single stage. Focus of this chapter is to consider and analyse manufacturing systems and their suppliers, because it is one of the most critical phases, concerning sustainability. Furthermore, LCC and LCA have been identified as the methodologies able to help decision makers in the improvement of industrial system's sustainability. However, the above methodologies have a series of limitations, which want to be covered by the closed loop framework proposed. Indeed, after the study and identification of the most critical phases within the industrial system life cycle, a closed loop framework, completed by tools, has been created and developed. Next step will regard the inclusion into the framework of the re-design and End of Life phase, in order to cover the whole life cycle for an efficient and effective management of industrial systems.

5.4 Workplaces of the Future and IoT: The FITMAN Project

The mission of the FITMAN (Future Internet Technologies for MANufacturing industries [20]) project is to provide the Future Internet Public Private Partnership with 10 industry-led use case trials in the domains of Smart, Digital and Virtual Factories of the Future.

The project aims to test and assess the suitability, openness and flexibility of FIWARE [21] OSS generic components (GEs) and provide FITMAN OSS specific components for manufacturing (SEs). The interested reader is referred to [22, 23] for a more detailed introduction to the FITMAN IoT approach.

The aim of this section is to provide a deeper insight about FITMAN project regarding the "IoT for Manufacturing" trials and requirements, exploitation plans and business opportunities.

5.4.1 FITMAN Smart Factory Platform (IoT)

FITMAN provides three Reference Platforms that conceptually describe the interconnection between the OS components on a specific Factory of the Future domain (smart, digital and virtual).

The concept of FITMAN Platform is a blueprint that allows the development of value added services in the smart-IoT, digital information management and collaboration areas.

More concretely, the Smart Factory Platform provides added functionalities to build innovative IoT services on top of FIWARE and FITMAN OSS components (see Figure 5.5).

This Platform is focused on the collection and processing of real-time data collection for shop floor operations. It uses sensors and monitoring systems

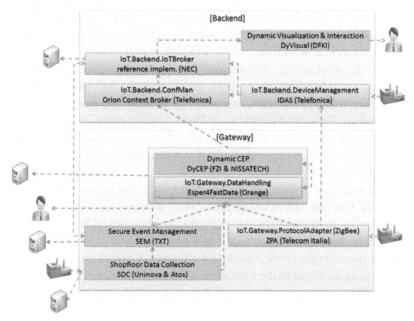

Figure 5.5 FITMAN Smart Factory Platform for IoT Services.

(Kinect sensor, RFID sensors, etc.) deployed in the production lines and services/applications for sending messages or warnings.

It aims to manage both tangible assets (energy, productivity, throughput, and wastes) and intangible assets (customer sentiments, workforce wellness, comfort and safety) to obtain more competitive and productive manufacturing environment.

As a result, more efficient knowledge-management strategies will be implemented, achieving the intelligent and smart factory.

5.4.2 Safe & Healthy Workforce: TRW Use Case

Active and healthy ageing is one of the major societal concerns in the smart home, smart city and smart industry contexts. In fact, human and organi-sational factors are still involved in almost 90% of all workplace accidents and incidents. The European Commission estimates that Musculo-Skeletal Disorders (MSD) accounts for 50% of all absences from work lasting 3 days or longer and for 60% of permanent work incapacity. Furthermore, up to 80% of the adult population will be affected by an MSD at some time in their life. In the last few years, it has become apparent that this situation will even worsen with the emergence of a European aging workforce [24], which may exhibit more limited physical and cognitive responses. According to the World Health Organisation, European workforce will be older than ever before, it will make up for 30% of the working-age population [25].

For this reason, it becomes necessary to develop new solutions that, taking benefit from the IoT technologies, improve the quality and comfort of the workforce. Indeed, TRW trial (worldwide Tier 1 – automotive supplier) aims to develop a new generation of worker-centric safety management systems in order to reduce accidents and incidents in the production workplace through workers' empowerment.

The TRW use case is focused on the monitoring and assessment of the ergonomic risk that can affect to the blue-collar workers on the production lines, in order to perform effective prevention strategies. The traditional prevention strategies are not capable of customizing specific plans and current human-based surveillance is not completely effective. Current procedures and systems are not customized to the limitations or characteristics of the workers, so the results are not trustworthy in all the cases. Thus, the trial specially provides the following functionalities: i) real-time data collection through Kinect sensors (see Figure 5.6a), ii) continuous data processing for ergonomic risks detection, iii) events management, and iv) web services and applications for corrective actions performance (e.g. web services for sending messages or warnings related to the risks detected; see Figure 5.6b).

Figure 5.6 TRW trial real environment (a) and GUI (b).

The TRW Trial Platform (see Figure 5.7) is based on the Smart Factory Platform (GEs and SEs), as well as TRW legacy system, and new components and services developed for the trial (TSC). The TRW Trial Platform exploits the OSS components mainly related to IoT services, taking advantage of functionalities related to data gathering, complex event processing, context information management, and event information delivery, among others.

As a result, the system monitors and collects real-time data of the skeleton and movements of the workers in the production line through Kinect sensors. It processes the data in order to detect "unsafety" events based on the defined rules (e.g. angles of the body or frequency of movements higher than the threshold values). In case of event detection, the different services or actions previously set up are delivered, e.g. messages or warnings are sent to the blue-collar workers and prevention technicians.

Preliminary intermediate KPIs already demonstrate a reduction of 13% in the number of accidents and incidents in the factory, as well as a performance improvement of 80% in the number of risks detected and alarms activated.

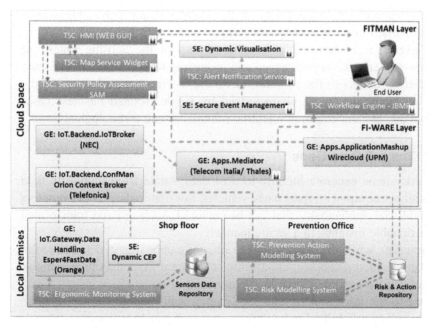

Figure 5.7   Future Internet Platform for Safer and Healthier Workplace.

TRW trial supports the introduction and acceptance of IoT technologies in the manufacturing industry in order to get the workplace of the future, providing the necessary balance between security concerns and privacy concerns. The implementation of monitoring technologies supports the innovative human-in-the-loop model, getting a participative approach for the workers' empowerment solution.

5.5 Osmosis Membranes for IoT Real-Digital-Virtual Worlds Interconnection: The OSMOSE Project

According to the FInES Research Roadmap 2025, Sensing Enterprise and Liquid Enterprise are two Qualities of Being, which are considered strategic for any future enterprise.

The Sensing Enterprise will emerge with the evolution of the Internet of Things, when objects, equipment, and technological infrastructures will exhibit advanced networking and processing capabilities, actively cooperating to form a sort of 'nervous system' within the enterprise next generation.

Among the challenges for applying IoT to the manufacturing enterprises in several domains, the Predictive Maintenance and the Asset Management applications will increasingly rely on Big Data analysis, Trust and Security system hardening technologies and Global Service technologies. From all these events, it is quite clear that the take-up of the Sensing Enterprise concept will enable very advanced and promising new business models and applications thanks to the adoption of Future Internet technologies, but the depicted landscape seems characterized by single isolated business cases without any evident common scientific framework and technological base.

5.5.1 The IoT Data Gaps

IoT ecosystems generate literally tons of data – data from mobile in-field applications, equipment, spare parts, vehicles, raw materials, process monitors and industrial appliances – that can be captured, analyzed and transformed into actionable insights, in a secure manner.

However, while the real world data streams can be ingested and stored in fast-data enabled databases, not all information are immediately relevant to support decision making, or action optimization.

Due to the large size of data sets, even a sanity check on data quality, coherency and reliability is a hard, time-consuming task. Furthermore, trends and deviations from expected systems' behaviours require accurate model building and checking while refining the model itself in real time. A digital world is then superimposed to the real one containing the computed data (performance trends, quantitative process and product metrics, tooling configurations, validation test patterns and results, etc.) and the data-driven decisions are taken, impacting on both real and digital worlds.

Thanks to the simulators, we are now able to create virtual data from model themselves under a set of stimuli, and compare those virtual behaviours with the real ones. A new stream of data is now generated from the Virtual world by interaction with virtual smart objects, impacting again on both real and digital worlds. The present state of art reports the existence of IoT data gaps among those worlds, exposing an intrinsic weakness in building a common view crossing the three worlds due to lacking of unified representation and data management tools.

5.5.2 The Liquid Enterprise

The Liquid Enterprise is an enterprise having fuzzy boundaries, in terms of human resources, markets, products and processes. Its strategies and operational models will make it difficult to distinguish the 'inside' and the 'outside'. Every activity in the enterprise is enabled by IoT devices and it streams data.

The intuition of the OSMOSE project is to explain the sensing enterprise by means of a metaphor taken form physics and strongly supported by the Liquid Enterprise idea.

Let us in fact imagine the Sensing-Liquid Enterprise as a pot internally subdivided into three sectors by means of three membranes and forming the Real-Digital-Virtual sectors. A blue liquid is poured into the first sector (Real World population), a red liquid into the second sector (Digital World population) and a green liquid into the third sector (Virtual World population).

If the membranes are semi-permeable, by following the rules of osmosis which characterizes each of the three membranes, the liquid particles could pass through them and influence the neighbouring world, so that in reality in the blue Real World we could also have red-green shadow ambassadors of the Digital/Virtual World and similarly for the other Worlds.

An entity (a person, a sensor network, an intelligent object) in the blue Real World could have control of their shadow images in the red Digital World (digital twin) and in the green Virtual World, keep them consistent and pass them just the needed information under pre-defined but flexible privacy and security policies.

The liquid enterprise paradigm has been considered so far just as a provocation, with very scarce understanding and adoption by industry and manufacturing enterprises in particular.

5.5.3 Osmotic Context Management

In Sensing/Liquid Enterprises, shadow images of entities are present in the real, virtual and digital world. Semi-permeable membranes manage the data flow between the worlds and thus keep background consistency. Thus, context awareness and understanding the meaning behind data is a major challenge. Context is concerned with relevant characteristics that describe an entity's situation [13]. In Sensing/Liquid Enterprises, these characteristics are spread between the worlds, posing new challenges for context management.

Knowledge need to be structured and linked appropriately. Ontologies enable machine-readable knowledge structures shared by the three worlds. Ontology modularization [14] allows detachment of horizontal (entities, processes, services, etc.) or vertical (entities in the real, digital or virtual world) domain ontology modules. Upper-Ontology modules provide generic knowledge structures that domain ontology modules have to obey. Knowledge Links [11, 12] describe interrelations between ontology modules.

Figure 5.8 illustrates the modular ontology approach of the OSMOSE Project, where entities, processes, services and events as well as platform

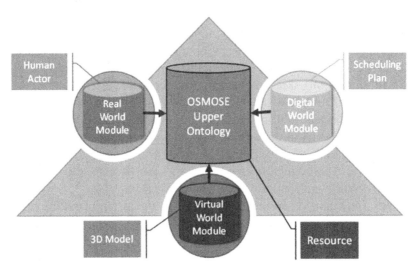

Figure 5.8 OSMOSE Ontology Approach.

specific terms (e.g. osmotic process) are defined and horizontally detached in the OSMOSE Upper Ontology, inherited by the three worlds and further specified by each of the world inside the vertical detached world ontology module.

5.6 IoT Enhanced Learning for Complex Systems Maintenance: The TELLME Project

The TELLME project (www.tellme-ip.eu) aims at providing a new methodology and an ICT technical architecture to provide personalized learning to technicians at the time they are working on complex systems. The training is continuously tailored on the activities that the worker have to accomplish (the daily schedule), the worker profile (ability on certain required operation) and the context in which he is immersed. Starting from those information the core of the TELLME system is able to support the execution of a job (usually described in a jobcard or workcard) providing to the worker a personalized and contextualized training. It is based on briefing (learning to be done before to the job), support (learning by doing and contextual learning contents) and debriefing (training to be done after the work to optimize the training). By its nature TELLME system should be fully aware of the context in which the user is immersed, the advancements status of the work, the respect of the rules of the workplace and, of course the safety. The usage of Internet of Things is essential for TELLME in order to tune the training for the worker.

Here below the need for IoT enhanced Learning is detailed for a specific complex system: helicopters maintenance. After that, two scenarios of IOT enhanced learning are provided. In the first one, the worker and the optimization of the training uses the IOT to monitor the compliance of worker activities with safety rules detecting and correcting potential dangerous behavior. The second scenario monitors the environmental conditions in order to keep the worker aware of them and providing specific learning about how to fine tune the activities to react to non-standard conditions.

5.6.1 The Need for IoT Enhanced Learning in Aerospace

People who perform aviation maintenance are unique. While all maintenance technicians learn their job skills through a combination of education, formal training, and "hands on" experience, aviation maintenance training has a special focus and unique challenges. This is due to stringent safety and quality

Accidents and incidents related to human errors vs. mechanical or other failures

▪ Accidents and incidents related to human errors

▪ Accidents and incidents related to mechanical or other failures

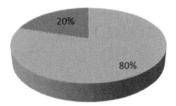

Figure 5.9   Accidents and incidents related to human errors vs. other failures.

requirements, complex equipment with sophisticated systems and a wide range of working conditions.

However aeronautical accidents and incidents are more likely to be caused by the actions of humans than by mechanical failure. Industry statistics show that human error contributes to nearly 80% of airline accidents and incidents, as illustrated below. This figure includes all aspects of human factors including operations, maintenance, and air traffic control. Because aviation systems are continually improving, the aircraft is seldom the cause of an accident and incident. Humans, rather than equipment, are more likely to be at the root cause of an accident or incident. Therefore, the best opportunity for safety improvement is to understand and manage the human factors that pose safety risks.

The investigation highlights a variety of maintenance factors that contributed to the incident, many of which were rooted in human factors. The factors also included the following:

1. inadequate training of technicians;
2. inadequate environment condition in the maintenance workplace conducing to error;
3. lack in the management of FOD-Foreign Object Debris.

In order to face these issues, over the past several year, a greater focus is placed on how human factors for improving aviation safety. First of all human factor training for aviation maintenance organizations is a mandatory requirement by international civil aviation authorities by means of "formal courses" in which technicians must attend to obtain certification that allows them to work as well as with continuous training (refresh courses) and at the workplace support. With regard to points 2 and 3 of the above list, the technology and the Internet of Things can effectively contribute to the reduction of risks.

5.6.2 IoT Enhanced Learning for Avoidance of Foreign Object Debris (FOD)

Foreign Objects are a major cause of aircraft damage and unscheduled maintenance. Damage can result in minor repairs or catastrophic events. Preventing FOD is everyone's responsibility. FOD includes hardware, tools, parts, metal shavings, broken hardware parts, pavement fragments, rocks, badges, hats, paper clips, rags, trash, paperwork and even wildlife. Any foreign object that can find its way into an aircraft or engine can contribute to FOD. FOD prevention is an essential element in all maintenance activities and is the responsibility of every company technician. About tools monitoring there are numerous methods to facilitate accountability of tools (screwdriver, torque wrench, rivet gun, air hammer, clecos, etc.). These include but are not limited to the use of tool inventory lists, shadow boards, shadowboxing, bar coding, special canvas layouts with tool pockets, tool counters, chit system, tool tags, or consolidated tool kits. In the last years the main manufacturers of aeronautical tools started developing electronic tool control based on smart tool box which records exactly who enters the box and when using an electronic badge, keyless entry and which tools have been removed or returned, and then confirm which tools are being issued and/or returned.

Inside the AgustaWestland organization, the workers involve in helicopter final assembly or maintenance activities are properly committed in FOD prevention best practices. Moreover specific processes and procedures have been put in place to prevent potential risk situations and to encourage workers to a positive attitude towards safety and to discourage wrong behaviour.

The standards achieved in FOD prevention by AW are at the highest levels; nevertheless it is important to maintain a high level of attention and put in place every resource that can contribute to reduce risks of FOD. For this reason AgustaWestland is experimenting with TELLME a new Internet of Things based monitoring of workers behaviour and FOD learning. First of all AgustaWestland is equipping its workers with smart tool boxes able to trace when the tools are taken and when they returned and those data are exported and used by the system to infers the workers behaviour and start corrections is necessary.

Figure 5.10   Human involvement in assembly process.

By this approach after the usual training workers, during daily worker, are monitored to detect potential FOD BEFORE they happen, in case of FOD identify the causes (motivation, knowledge or structure) and in case of no behaviour changing personalized (optimized) training based on B-BS (Behavior-Based Safety) methodology.

The architecture of the application is reported in the picture below. The IOT based solution is implemented by components providing events about what is happening, a second layer of components managing events and detecting potential behaviour and a third level managing learning contents to correct behaviour.

In the workplace layer, during the daily work, the worker should respect a specific pattern in all maintenance activities: start maintenance activity, get tools, return tools and close maintenance activity, not respecting it means that a tools could have been left in the helicopter. The Platform monitors the environment in order to discriminate which BCW (if any) needs to refresh their tool management to improve the performances. Events about tools (get and return) from the smart toolbox (SnapOn) and maintenance activities (start-close) from the mobile electronic work-card are gathered by the adapters and published to a queue of the event broker.

In the event management layer events aggregation is realised by a set of adapters developed specifically for the different event sources and sending them in AMQP (Advanced Message Queuing Protocol) protocol. Events are managed by an event Broker implemented by RabbitMQ aggregating them

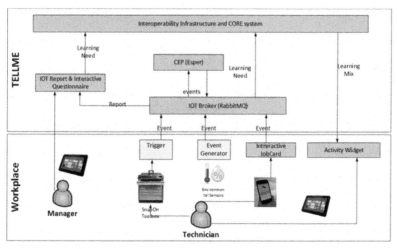

Figure 5.11   TELLME Approach.

and collaborating, for the control of the pattern, with the Complex Event
Processor implemented with Esper. In case of non-respect for the pattern an
additional event (learning need) is generated triggering a BPEL (Business
Process Execution Language) process in the interoperability infrastructure to
assign a focused lesson to fill the skill gap of the worker. If the information
about the worker are enough to take an automatic decision about which
learning contents to provide to him a learning mix (set of learning contents) is
created and the reference of it is notified to the user mobile device. In case the
automatic decision cannot be taken (eg: there are contradictions) the manager
of the technicians is notified and a human observation required. It is essential,
in fact, to understand if the wrong behaviours is due to a problem of knowledge
(the worker didn't know to do something) or a problem of motivation (he
knows but was not willing to) or a problem of structure (someone stole him
the tools for another job). The observation and the interview the manager has
to do provides more information that makes the system able to take a decision
and provide the right mix of learning contents to modify the behaviour.

5.6.3 IoT Enhanced Learning for Non-Standard Workplace Environmental Condition

Workplace environmental conditions can impact the quality of work performance and technician fatigue as well as changing a bit processes for
some specific steps. However, each day aviation maintenance workers are

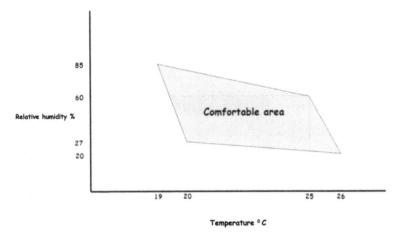

Figure 5.12   Environmental condition for comfort.

sometimes faced with sub-optimal work conditions which contribute to fatigue. When these conditions can be controlled they must be. In 1989, the National Transportation Safety Board issued recommendations urging all modes of transportation to conduct research on fatigue and on the workplace environmental condition can affect it. As results of these researches in 1999 the FAA (Federal Aviation Administration) issued a CD-ROM titled Human Factors in Aviation Maintenance and Inspection (HFAMI) collecting various types of documents such as conference proceedings, research reports, and the "Human Factors Guide for Aviation Maintenance." In which the "Chapter 3 – Workplace Safety" states how a hangar floor should be organized.

In particular, a rage of humidity, temperature and light has been fixed to create a "comfortable area" of work; adequate light is ensured by the large glazed areas of the hangar roof and complemented by fluorescent tubes up to 300 lux illumination. Even manual operations, in case must be done even if outside the comfortable area should be adapted like in the case of using grease, under a certain temperature should be managed in a slightly different way with more care than normal.

The goal of TELLME pilot is twofold: make the worker perfectly aware of conditions in order to avoid any problems and provide tailored training contents to train maintenance technicians in working on borderline conditions (e.g., how to manage grease). The solution implemented is of course leveraging on the Internet of Things; the workplace has been equipped with sensors for continuous monitoring of the environmental and able to notify a warning to the worker if the conditions (being out of comfortable area) could lead to safety risks for the worker and for the helicopter. Technically, the prototype has been implemented by a RaspberryPI on which humidity, light and temperature sensors are integrated and the final box positioned close to the workers. Different kind of messages (e.g., "over threshold", "still over threshold") are sent to the worker by its own mobile devices used to run interactive jobcards. Messages are sent from raspberry by AMQP protocol, received by RabbitMQ as broker a managed by a complex event processor implemented by Esper controlling whether or not the comfortable zone is respected and which are the implication with the current work-card describing the maintenance to be executed. To make the worker aware about the working conditions, warnings are sent by a web service as push messages and visualized by an Android native application on worker's smartphone. At the same time the warning are matched with the work card by the system that retrieves the right learning contents for the specific environmental conditions and push them to the workers by a specific GUI on the mobile device of the worker.

5.6.4 Future Work

The two scenarios are just few of the IoT enhanced Learning for complex systems maintenance that can be realized. TELLME is running the second iteration of the pilot in order to validate the solution. Firsts feedbacks proven the feasibility of the solutions and the positive impact especially in terms of worker stress; especially the FOD is a very important and stressful element of the daily work in aerospace and the IoT system designed is well seen by workers and an important support.

5.7 IoT-Driven Manufacturing Innovation Ecosystems

Innovation is one of the factors underpinning the success of manufacturing in current global markets. In a world where customer demands change at an accelerated pace and competition is ever growing in numbers and intensity, innovation is essential for business to remain productive and successful. This applies not only for New Product Introduction (NPI) activities but also to innovation in processes, equipment, ICT and other enabling aspects of manufacturing. In spite of its importance, innovation is still seen as a major challenge in manufacturing. Even though innovation usually requires information and support coming from an interactive ecosystem, the process still seems to be lineal, uninformed, slow and segregated. Given the nature of the limitations, the Internet of Things (IoT) has huge potential to ***drive innovation by connecting manufacturing ecosystems***.

The key to the potential role of IoT as a driver for manufacturing innovation is in the nature, extension and diversity of innovation cycles in manufacturing. Innovation cycles usually start with a new idea for a manufacturing entity, whether is a product, a process or a machine. This is followed by definition, prototyping and test stages, after which the final version is generated. Finally, there are steps to manage the in-life and end-of-life for the entity.

In manufacturing, several lifecycles coexist and converge during operation (Constantinescu, Landherr, & Neumann, 2013). The most important lifecycles are for product and operations. However, other dominant lifecycles are for factories, ICT services and equipment (an example of the convergence of these cycles, adapted from Constantinescu, et. at. (2013), is shown in the following picture Figure 5.13.

Innovation cycles rely on key feedback loops and interactions with internal and external actors. For example, a new automotive engine would be designed taking into consideration warranty information from previous models to avoid known issues and customer dissatisfaction. Historic information may be pulled from internal databases to understand the causes of the most significant issues, enabling design decisions for optimized products. This kind of feedback loop

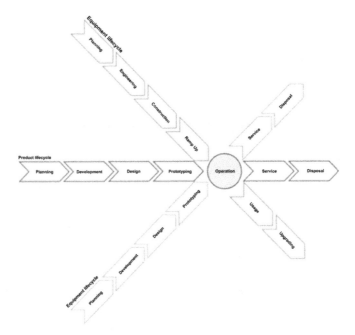

Figure 5.13   Convergence of lifecycles.

requires interactions and information exchanges between different actors and objects in the product lifecycle. In this case, the engine in service and the customer would interact with the designer and the design being generated. This is a typical example of an interaction that would accelerate and optimize product innovation. Other interactions occur between different lifecycles. One such example would be the interaction between an equipment designer and an organization that uses the equipment to manufacture consumer products. In this case, the equipment designer may use performance and maintenance records to generate and improve design. Finally, there are interactions between producers and supporting actors such as government or R&D organizations.

Several solutions have been developed in the past to support these interactions. In the case of interactions within the product lifecycle, Product Lifecycle Management (PLM) systems and associated tools have offered substantial support during the past decades. As for cross cycle interactions, organizations have moved towards remote monitoring of their products and a Product Service System (PSS) approach (Mont, 2002). However, interactions are not always straightforward, effective or even possible due to connectivity issues.

The Internet of Things (IoT) is the concept of internet-enabled "things" (e.g. machines, sensors and mobile devices) to facilitate exchange of relevant information that can be ultimately used to create benefits for an individual or an organization. The concept has already been developed successfully in the consumer world. In the industrial context, it is already starting to show potential to drive smart manufacturing by facilitating, for example, the exchange of information between products and machines to enable self-adaptive processes and operations. As an enabler of connectivity, *IoT also holds huge potential to drive innovation in manufacturing by forming real manufacturing innovation ecosystems*, facilitating interactions between the relevant actors and lifecycles.

The idea of seamless acquisition and transfer of the right information, to the right place, at the right time to drive innovation and the possibility to link the information to applications and people to draw further support, is a powerful one. IoT can enable the involvement of a diverse and heterogeneous ecosystem of organizations to exchange information and ideas for innovation. The support can be provided mainly at three levels, as listed below.

- **Single innovation cycle connectivity**: This is where IoT is used to create connectivity between "things" within a single innovation cycle. The most representative example of this level would be the product lifecycle. In that case, IoT would enable products to be tracked throughout their whole life, absorbing relevant information and feeding it to design actors at the beginning of the cycle, involving the whole supply chain.

- **Cross-cycle connectivity**: At this level, the word "ecosystem" become more meaningful, as it brings up collaboration between diverse actors in a manufacturing environment. For example, an Original Equipment Manufacturers (OEM) could perform design based on information from their own equipment in the equipment lifecycle, but also include the impact of the equipment on product quality by adding connectivity to "things" in the product lifecycle. At this level, interactions become complex and security and privacy become prime concerns.

- **External connectivity**: Interactions between organizations that are not necessarily part of a manufacturing ecosystem or even the manufacturing sector may be valuable. The main interactions in this area are with organizations in the innovation chain (e.g. universities, research institutes, startups) and support networks (e.g. government and trade associations). Seamless connectivity with this kind of partner would bring additional knowledge and services to further accelerate innovation.

In summary, the vision of IoT-driven Manufacturing Innovation Ecosystems is one where all "things" (products, machines, tools, sensors, mobile devices) are connected through the internet; where design-relevant information is captured and transferred to the right actors or applications; where applications that orchestrate the right interactions to drive innovation are enabled through an IoT network; where supporting organizations can provide information and services to accelerate innovation; and where transparency, collaboration and ownership are managed appropriately.

Such a vision can lead to Manufacturing Innovation Ecosystems that can anticipate new design requirements and react in an agile way to deliver novel solutions seamlessly. Ultimately, such ecosystems would thrive to capture market share by providing the best products produced by the most advanced manufacturing systems.

This section illustrates the potential of IoT to drive manufacturing innovation ecosystems and highlights the importance of adopting the approach. Furthermore, IoT as a driver of innovation in manufacturing opens the field for new opportunities that could lead to significant business benefits for the sector. Some of the most significant opportunities are:

- Servitization of innovation lifecycle management systems (such as PLM and asset management systems) and creation of information driven innovation networks around them;

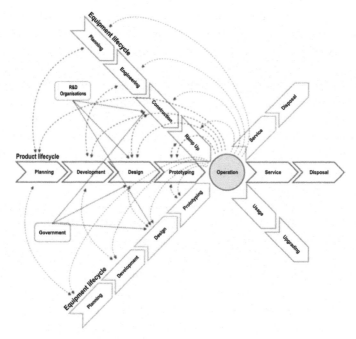

Figure 5.14   Innovation lifecycles interactions.

- Creation of functional and support partnerships to create strong pathways for accelerated innovation driven by IoT and the mechanisms required for smooth operation (e.g. security, privacy, legislation, contractual agreements);
- Active co-creation or collaborative innovation, where partnerships, networks or communities connect to create a new solution jointly based on the best information, knowledge and skills available;
- Enabling open innovation (Chesbrough, 2003) by facilitating authorized and agreed exchange of knowledge between organizations for the purpose of innovation, increasing levels of collaboration, opportunities for disruptive innovation and new business models.

5.8 Industrial Internet of Things: The US IGNITE Perspective

This section provides an overview of technologies, tools and practices pertaining to Internet of Things (IoT) in context of the US Ignite initiative. An introduction to IoT is first provided along with the objectives and aims of the US Ignite initiative. This is followed by a discussion of cyber physical approaches and frameworks and their impact on IoT. Finally, a discussion of a broad framework for collaboration using IoT and cyber physical frameworks is outlined.

The emphasis on adoption of IoT principles can also set in motion the realization of advanced next generation Cyber Physical relationships and frameworks which can enable software tools to control and accomplish various mundane as well as advanced physical activities; these applications can be a software entity on a smart phone controlling a robot helping a disabled person to drink a cup of water or it can be an advanced simulator being used as an interface to perform life saving medical surgeries in a hospital.

In an IoT context, one of the core benefits is from the cyber physical interactions which help facilitate changes in the physical world. The plethora of smart devices emerging in the market serves as a catalyst for this next revolution which will greatly impact manufacturing and manufacturing practices globally. Imagine being able to design, simulate and build a customized cell phone from a beach thousands of kilometers away from an engineering organization. Today, using cloud technologies and thin clients such as smart phones and smart watches, the potential of using such IoT principles and technologies for advanced manufacturing is very high.

Such cyber physical approaches also support an agile strategy which can enable organizations functioning as Virtual Enterprise partners to respond to changing customer requirements and produce a range of manufactured goods. With the help of advanced computer networks, such cyber (or software) resources and tools can be integrated with physical resources including manufacturing equipment. When customer requirements change, such an approach can also help interfacing and integrating with a variety of distributed physical equipment whose capabilities can meet the engineering requirements based on the changing product design. Against this backdrop, it is important to also underscore recent efforts to develop the next generation of Internets.

5.8.1 Background on US IGNITE and the GENI/FIRE Initiatives

In the US, the Global Environment for Network Innovations (GENI) is a National Science Foundation led initiative focused on the design of the next generation of Internets including the deployment of software designed networks (SDN) and cloud based technologies. GENI can also be viewed as a virtual laboratory at the frontiers of network science and engineering for exploring Future Internet architectures and applications at-scale.

In the context of advanced manufacturing (such as micro assembly), such networks will enable distributed VE partners to exchange high bandwidth graphic rich data (such as the simulation of assembly alternatives, design of process layouts, analysis of potential assembly problems as well as monitoring the physical accomplishment of target assembly plans).

In the European Union (EU) and Japan (as well as other countries), similar initiatives have also been initiated; in the EU, the Future Internet Research and Experimentation Initiative (FIRE) is investigating and experimentally validating highly innovative ideas for new networking and service paradigms (http://www.ict-fire.eu/home.html).

Another important initiative is the US Ignite (http://us-ignite.org/) which seeks to foster the creation of next-generation Internet applications that provide transformative public benefit using ultrafast high gigabit networks. The six US national priority areas are the following:

- Health
- Advanced manufacturing
- Public Safety
- Education & Workforce
- Energy
- Transportation

Both these initiatives herald the emergence of the next generation computing frameworks which in turn have set in motion the next Information Centric revolution in a wide range of industrial domains from engineering to public transport. These applications along with the cyber technologies are expected to impact global practices in a phenomenal manner. As research in the design of the next generation Internets evolves, such cyber physical frameworks will become more commonplace. Initiatives such as GENI and US Ignite are beginning to focus on such next generation computer networking technologies which hold the potential to radically change the face of advanced manufacturing and engineering (among other domains).

5.8.2 Cyber Physical Tools and Frameworks

IoT entities and devices will greatly benefit from the evolution of Cyber Physical approaches, systems and technologies. The term 'cyber' can refers to a software entity embedded in a thin client or smart device. A cyber physical system can be viewed as an advanced collaborative collection of both software and physical entities which share data, information and knowledge to achieve a function (which can be technical, service or social in nature).

In a process engineering context, such cyber physical systems can be viewed as an emerging trend where software tools can interface or interact with physical devices to accomplish a variety of activities ranging from sensing, monitoring to advanced assembly and manufacturing.

In today's network oriented technology context, such software tools and resources can interact with physical components through local area networks or through the ubiquitous Word Wide Web (or the Internet). With the advent of the Next Generation Internet(s), the potential of adopting cyber physical technologies and frameworks for a range of process has increased phenomenally.

In the context of manufacturing, collaborations within an IoT context can be realized using various networking technologies including cloud based computing. According to the National Institute of Science and Technology (NIST), Cloud computing can be viewed as a model for enabling ubiquitous, convenient, on-demand network access to a shared pool of configurable computing resources (including networks, storage, services and servers) [29]; the computing resources can be rapidly provisioned with reduced or minimal management effort or interaction with service providers. Some of the benefits for cloud based manufacturing include reducing up-front investments and lower entry cost (for small businesses), reduced infrastructure costs, and reduced maintenance and upgrade costs [28]. In [27], Tao et al discussed the context of Internet of Things (IoT) and Cloud Computing (CC) which hold the potential to providing new methods for intelligent connections and efficient sharing of resources. The authors propose a service system which consists of CC and IOT based Cloud Manufacturing.

One of the US Ignite projects dealing with advanced manufacturing involves the creation of a GENI based cyber physical framework for advanced manufacturing [30]. The manufacturing domain is the assembly of extremely tiny micron sized devices. This is the first research project involving Digital Manufacturing, cyber physical frameworks and the emerging Next Generation Internet (being built as part of the GENI initiatives). A cyber physical test bed is under development to enable globally distributed software and manufacturing resources to be accessed from different locations and used to accomplish a complex set of life cycle activities including design analysis, assembly planning, simulation and finally assembly of micro devices. The presence of ultra-fast high gigabit networks enables the exchange of high definition graphics (in the Virtual Reality based simulation environments) and the camera monitoring data (of the various complex micro manipulation and assembly tasks by advanced robots and controllers). Engineers from different locations interact more effectively when using such Virtual Assembly Analysis environments and comparing assembly and gripping alternatives prior to physical assembly. GENI and FIRE Next Generation technologies also embrace software defined networking (SDN) principles, which not only reduces the complexity seen in today's networks, but also helps Cloud service providers host millions of virtual networks without the need for common separation isolation methods such as VLAN [31]. SDN also enables the management of network services from a central management tool by virtualizing physical network connectivity into logical network connectivity [31].

The US Ignite initiative focuses on the creation and fostering of 'transformative, next-generation applications' [32]; it seeks to create new business opportunities that will accelerate U.S. leadership in the adoption of ultra-fast broadband and software-defined networks both within and outside the US. These applications will range from how doctors get trained as well as provide medical services to how to build products faster and at a lower cost.

5.9 Research, Innovation Challenges for IoT Adoption in Manufacturing: The SMART 2013/37 EC Study

The SMART 2013/37 EC study, entrusted by the EC DG CONNECT to IDC EMEA and TXT e-solutions, investigates the enabling factors for the European industry in the market emerging from the combination of IoT, Cloud Computing and Big Data, and provides a set of recommendations to foster European research activities and its capability to catch new market opportunities.

The study initially provides an overview of the IoT ecosystem in Europe and sketches possible evolutions of this ecosystem in 2020. Afterwards it tries to identify emerging markets and new business opportunities the mixing of IoT and Cloud Computing can offer within in the coming years. Finally, the study identifies the related research and innovation challenges and devises a coherent strategy and actions to address the identified challenges and foster the take-up of the IoT market in Europe and strengthen European actors' position on this market. Even if the study was not specifically focused on the Industrial IoT, due to the relevance of manufacturing in the European market and the leading position of Europe in the manufacturing area, the study provides useful elements to frame IoT within manufacturing.

5.9.1 The Study IoT and Cloud Research and Innovation Strategy

The disruptive nature of IoT, Cloud Computing and Big data, which is further enhanced by their combination, require a suitable methodology able to combine the traditional technology push and demand pull forces and to take into account the various relations and the potential stakeholders. The adopted guiding methodology to devise a Research and Innovation Strategy is depicted in Figure 5.15.

As clear from the figure, the methodology combines a forward analysis path (the upper red arrow) with a backward analysis one (the lower red arrow). The forward path is in line with the traditional technology-driven approach where research challenges spur the development of new technologies that bring out new business models and new market opportunities.

The backward analysis stream, instead, starts from the identification of new potential market scenarios and key factors to identify the main demand needs and requirements in the research area. The objective of the study research and innovation analysis methodology is to match requirements to research challenges, align the demand needs with the innovation challenges and foster the take-up of the most promising scenarios.

5.9.2 The Main Market Trends

The market analysis moved from framing the IoT Ecosystem (see Figure 5.16) and the identification of the main interaction patterns.

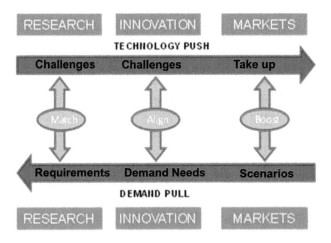

Figure 5.15   Conceptual Model of the development of the R&I strategy.

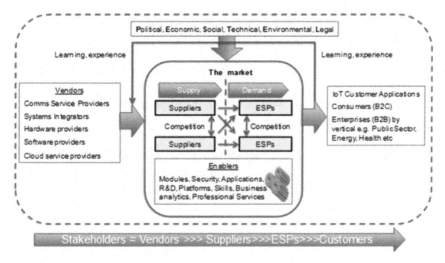

Figure 5.16 The IoT ecosystem.

The IoT ecosystem is heavily dynamic and is still evolving. For example, currently there is still a relevance of vertical markets, but the deployment of open standards and horizontal platforms will quickly change this aspect.

The main stakeholders' categories identified for the IoT ecosystem can be summarised as follows:

- **Vendors** that supply components to the solution providers. This category encompasses a large variety of enterprises for dimensions and specializations, ranging from large global multinationals to SMEs;
- **Suppliers** who develop IoT solutions or provide IoT related services;
- **Customers/end-users** who use IoT solutions or services.

On the main trends side the study splits them according to the analysis methodology described above into:

- **Technology push trends**:
  - *Enhanced connectivity infrastructure*: the availability of suitable network connectivity is, on the one hand, a pre-requisite for IoT take-up, and, on the other hand, the increasing anything, anywhere and anytime connectivity is boosting IoT;
  - *Cloud computing*: the availability of scalable and easily accessible virtualised resources is another key factor element;

- *Big Data*: also identified as *analytics* is another technological trend that is already characterizing the IoT market but will further characterize it in the near future. *Analytics* will span from the edge (i.e., *Fog/Edge Computing*) to the platforms on the cloud;
- *Increasing role of smart devices*: local sensing, computing and communication features will characterize all products in the coming years;
- *Horizontal Platforms*: service providers will push the market to adopt horizontal platforms therefore favouring economies of scale and the development of core services spanning different market segments.

- **Demand pull trends**:
  - *Demographic trends*: the aging population in Europe and Western World drives the need of IoT solutions especially in the healthcare and wellness market segments;
  - *Environmental consciousness*: the development of a culture of eco-consciousness will create a demand for IoT solutions able to support the reduction of the environmental footprint of our societies;
  - *Public Sector driving role*: the public sector will still play a role in asking for solutions able to improve their capabilities and efficiency in managing cities, transportation, health, tourism, etc.;
  - *Business demand/Consumers demand*: the business environment will ask for solutions able to improve their efficiency, expand their products' offering, increase the synergies and merging between products and services (servitization), while consumers will ask for more intelligent appliances (e.g., smart home, smart cars) and anywhere, anytime accessibility.

The vision of IoT in 2020 as identified by the IDC-TXT study is summarized in Figure 5.17.

The vision can be summarized by the following figures:

- the expectation is to have across the EU 28 an excess of 6 billion IoT devices generating revenues for more than 1.81 billion Euro;
- the IoT ecosystem will encompass not only the traditional supply-side actors but also an increasing number of businesses and organizations serving and using the IoT;
- the hyper-connected society will be an established reality by 2020.

Figure 5.17 The Vision of IoT in 2020.

5.9.3 The IoT and Cloud Research and Innovation Challenges

Moving from the analysis of the current IoT scenario and challenges, and taking into account innovations related to the Cloud Computing, Big Data and other Future Internet technologies, the study identified a set of **Grand Challenges** and their most relevant research topics to be addressed in the next research actions of the Horizon 2020 Programme. The *Grand Challenges* are:

- **Open Integrated Architecture**: the evolution is directed towards a 3rd generation platform concept and towards the integration and networking of platforms in a consistent and coherent integrated event driven and service oriented architecture. The most relevant research topics for this challenge are:

- *From Networks of Things to the Internet of Things*: promote the development of standardized interfaces, protocols and architecture to move from a "network of things" (e.g., siloed solutions, gateway-oriented approaches) to a solution where all things are directly accessible from any other point in the Internet;
- *IoT Integrated Development and Runtime Environment*: promote the development of new programming paradigms and run-time environments able to support the development of more responsive, distributed, dynamic and resilient systems;
- *IoT Systems Lifecycle Management*: promote the development of standardized management features able to face the IoT devices and services complexity, dynamics and heterogeneity.

- **End-to-End Connectivity**: support the development of, and deploy, protocols and solutions able to assure end-to-end connectivity of heterogeneous IoT related systems (e.g., move to IPv6, meta-information and knowledge representation and exchange so that systems can expose in a more meaningful way their characteristics and more efficiently locate and use data or services provided by other systems, new architectural approaches like: "fog computing", Named Data Networking, cognitive networks). Related research topics are:

  - *Systems-of-Systems Networking*: IoT requires new network approaches that permit easy connectivity, control and communications, and address IoT specific issues (e.g., scalability, manageability, resource usage, etc.);
  - *Knowledge-based IoT Networking*: upper layers interfaces and protocols able to use knowledge to address the IoT connectivity and interoperability issues;
  - *IoT Systems self-configuration*: features to support (auto) configuration and reconfiguration of entities and systems.

- **Security by Design**: assuring security and protection of the whole IoT system in a consistent and holistic approach along its whole life cycle. This challenge also accounts privacy management. Scalability of security (and privacy) solutions is also one of the issues to be addressed. The related research are:

- *IoT Compatible Security Techniques*: develop new encryption and security techniques (including secure identification, secure configuration, etc.) that can be usable on resource-constrained devices, and be scalable;
- *IoT systems Trust Management*: investigate and develop new scalable and usable solutions for privacy and security, including protocols and mechanisms able to represent and manage trust and trust;
- *Security and Privacy by Design*: introduce security and privacy by design approaches to make security and privacy an integral part of the systems.

- **Semantic-driven Analytics**: the advent of new technologies for Big (stream and historical) Data analysis, knowledge-based reasoning and advanced (including distributed and real time) decision making. The impacts IoT will have on network architecture and "consciousness" of entities requires the development of more advanced, reliable and privacy aware approaches and technologies to make possible for entities to learn, think, and understand both physical and social worlds. Related research Topics are:

  - *Knowledge Extraction*: develop techniques that convert raw data into usable knowledge;
  - *Semantic-based Data Quality*: develop techniques that improve the trust/confidence level of acquired;
  - *Distributed and autonomous reasoning*: develop protocols and platforms that support knowledge dissemination and distributed, autonomous reasoning.

5.9.4 Study Conclusions and EC Policy Recommendations

The study outcomes were articulated in a set of strategic actions and recommendations. The strategic actions are structured in three main pillars:

- **Europe needs to invest in the development of technologies** for the IoT, Cloud, and Big Data combination, able to manage complexity, provide scalability, guarantee usability and preserve privacy;
- **Europe must develop the supply ecosystem** and bridge the gap between research and market;
- **Europe must promote and support the take-up of IoT by user industries**, building the critical mass of users needed to encourage the investments needed for massive adoption;

- **Europe must create favourable framework conditions for the development of the IoT ecosystem** giving priority to the following main areas of action: developing the necessary skills, building trust in the emerging IoT economy, removing the regulatory barriers, and encouraging international cooperation.

The recommendation were split into different categories (e.g., Policy Recommendations to the EC, Recommendations to National Governments, Recommendation to Large Enterprises).

In the following only the most relevant ones for the IERC cluster book are reported:

Recommendation 1: raise the priority of IoT research in the H2020 Programme and prioritize the investment in the development of technologies for the IoT, Cloud, and Big Data combination, identified as the main priority to meet demand requirements in the period 2016–2018;

Recommendation 2: increase investments in the IoT and Cloud research and innovation area to support more innovation and take-up actions and accelerate the market and ecosystem development;

Recommendation 3: promote the development of broad-based, open horizontal platforms, in order to overcome the potential fragmentation of the EU market and to support the development of a competitive supply industry and a balanced ecosystem;

Recommendation 4: the EC should pay specific attention to the inclusion of innovative SMEs and start-ups in the research and innovation actions, incentivizing their active participation and making sure that they can access the necessary technology platforms to develop applications and services;

Recommendation 5: the EC should implement Large Scale Pilot (LSPs), or other innovation actions, in the most relevant emerging IoT markets (Smart Energy, Smart Transport, Smart Manufacturing, Smart Government, Smart Health);

Recommendation 6: the EC should promote the development of e-leadership skills for IoT;

Recommendation 7: the EC should contribute to the promotion of IoT readiness, by assessing the main techno-economic and network infrastructure requirements and the potential risks of digital divide across Europe;

Recommendation 8: the EC should promote IoT take-up and adoption by 1) promoting "leading by example" small innovation projects focused on proving the business case or developing innovative business models for the emerging IoT service economy 2) promoting the aggregation of demand and the use of pre-commercial procurement mechanisms in the public sector;

Recommendation 9: the EC should promote the development of communities of stakeholders;

Recommendation 10: the EC should ensure interoperability and security within Europe's major initiatives working closely with the Connecting Europe Facility (CEF) programme;

Recommendation 11: the EC should evaluate the possibility to launch an IoT and Cloud Public-Private Partnership (PPP) to manage more effectively innovation and take-up actions, and specifically an Accelerator programme to fund innovative start-ups and SMEs bringing to market new IoT-based products and services;

Recommendation 12: the EC should contribute to building trust and confidence in IoT by making sure that the research and innovation actions take into account psychological, social and pragmatic issues potentially affecting the trust and confidence of the potential users in IoT, Cloud, and Big Data solutions and services;

Recommendation 13: the EC should help developing the internal single market for IoT services and applications, by promoting the adoption of open standards and interoperable solutions across Europe.

Bibliography

[1] http://www.grantthornton.com/issues/library/survey-reports/food-and-beverage/2014/09-Why-sustainability-makes-business-sense-in-6-survey-findings.aspx (31-03-2015).
[2] http://thebuildnetwork.com/leadership/sustainability/ (27-04-2015).
[3] http://www.unep.org/pdf/UNEP_LifecycleInit_Dec_FINAL.pdf (31-03-2015).
[4] http://www.lifecycleinitiative.org/starting-life-cycle-thinking/benefits/ (27-04-2015).
[5] http://www.atkearney.com/paper/-/asset_publisher/dVxv4Hz2h8bS/content/sustainability-a-product-life-cycle-approach/10192 (27-04-2015).

[6] http://www.volkswagenag.com/content/vwcorp/content/en/the_group/st rategy.html (27-04-2015).

[7] http://www.bmwgroup.com/e/0_0_www_bmwgroup_com/verantwortung /kennzahlen_und_fakten/sustainable_value_report_2010/einzelne_kapite l_englisch/11670_SVR_2010_engl_Sustainable_Operations.pdf (27-04-2015).

[8] http://www.fcagroup.com/en-US/investor_relations/financial_informat ion_reports/annual_report/2014/FCA_2014_Annual_Report.pdf (27-04-2015).

[9] http://www.trade.gov/press/publications/newsletters/ita_1008/sustainabl e-mfg_1008.asp (28-04-2015).

[10] B.S. Blanchard. *Design To Cost, Life-Cycle Cost.* 1991 Tutorial Notes Annual Reliability and Maintainability Symposium, available from Evans Associates, 804 Vickers Avenue, Durham, NC 27701.

[11] S. Dowlatshahi. The role of logistics in concurrent engineering. *Journal of Production Economics,* 44 (3):189–199, 1996.

[12] A. S. Munro. Let's roast engineering sacred cows. *Machine Design,* 67(3):41, 1995.

[13] A. Sanders, and J. Klein. Systems Engineering Framework for Integrated Product and Industrial Design Including Trade Study Optimization. *Procedia Computer Science,* 8(1):413–419, 2012.

[14] G. Rebitzer, T. Ekvall, R. Frischknecht, D. Hunkeler, G. Norris, T. Rydberg, W.P. Schmidt, S. Suh, B.P. Weidema, D.W. Pennington. Life cycle assessment Part 1:Framework, goal and scope definition, inventory analysis, and applications. *Environment International,* 30(5):701–720, 2004.

[15] Society of Automotive Engineers (SAE). *Reliability and Maintainability: Guideline for Manufacturing Machinery and Equipment.* Available from M-110.2, Warrendale, PA, 1999.

[16] Scientific Applications International Corporation (SAIC). *Life Cycle Assessment: Principles and Practice.* Available from 11251 Roger Bacon Drive, Reston, VA 20190, 2006.

[17] International Organization for Standardization (ISO). *ISO 14040 Environmental Management – Life Cycle Assessment – Principles and Framework.* 1997.

[18] IFIP-IPAC Task Force. *GERAM Generalised Enterprise Reference Architecture and Methodology,* Version 1.6.1. 1998.

[19] A. J., Nebro, and J. J., Durillo. *jMetal 4.3 User Manual.* Available from Computer Science Department of the University of Malaga. 2013.

[20] "FITMAN," [Online]. Available: http://www.fitman-fi.eu/

[21] "FIWARE," [Online]. Available: http://www.fi-ware.org/

[22] M. Spirito and J. Sola, in *Internet of Things Applications – From Research and Innovation to Market Deployment*, River Publishers Series in Communications, 2014, pp. 243–281.

[23] O. Lazaro, S. Gusmeroli and M. Isaja, "FITMAN Future Internet Enablers for the Sensing Enterprise: A FIWARE Approach & Industrial Trialling," in *IWEI*, 2015.

[24] J. Ilmarinen, «Promoting active ageing in the workplace,» *European Agency for Safety and Health at Work (OSHA)*, 2012.

[25] European Commission, [En línea]. Available: http://ec.europa.eu/ economy_finance/articles/structural_reforms/2012-05-15_ageing_report_ en.htm

[26] http://en.wikipedia.org/wiki/Internet_of_Things

[27] Fei Tao, Ying Cheng, Li Da Xu, Lin Zhang, and Bo Hu Li. "CCIoT-CMfg: cloud computing and Internet of Things based cloud manufacturing service system." (2014): 1-1.

[28] Benefits of cloud computing, http://www.mbtmag.com/articles/2013/05/ how-manufacturers-can-benefit-cloud-computing

[29] http://csrc.nist.gov/publications/nistpubs/800-145/SP800-145.pdf

[30] Cecil, J., https://vrice.okstate.edu/content/gigabit-network-and-cyber-ph ysical-framework

[31] http://www.serverwatch.com/server-tutorials/eight-big-benefits-of-soft ware-defined-networking.html

[32] https://www.us-ignite.org/about/what-is-us-ignite/

[33] The study is available at: http://ec.europa.eu/digital-agenda/en/news/ definition-research-and-innovation-policy-leveraging-cloud-computing- and-iot-combination

6

Securing the Internet of Things – Security and Privacy in a Hyperconnected World

**Elias Z. Tragos[1], Henrich C. Pöhls[2], Ralf C. Staudemeyer[2],
Daniel Slamanig[3], Adam Kapovits[4], Santiago Suppan[5],
Alexandros Fragkiadakis[1], Gianmarco Baldini[6], Ricardo Neisse[6],
Peter Langendörfer[7], Zoya Dyka[7] and Christian Wittke[7]**

[1]FORTH, Greece
[2]University of Passau, Germany
[3]Technical University of Graz, Austria
[4]Eurescom, Germany
[5]Siemens AG, Germany
[6]European Commission, Joint Research Centre (JRC), Italy
[7]IHP, Germany

6.1 Introduction

The Internet of Things (IoT) introduces itself as a basic set of technological enablers to support the provision of innovative applications that can improve the quality of life of people and industrial productivity. IoT is increasingly supported by various stakeholders and market players that see clear business opportunities in this field. Cities have also identified the potential of IoT both for providing smart services to their citizens and for boosting the local economy by providing opportunities for new jobs and new businesses. Industry is considering IoT's adoption to drive Industry 4.0. All these are key reasons why IoT has attracted so much attention lately in both the research and the industrial world.

Main research areas in the IoT world until now included the development of technologies to efficiently interconnect large numbers of devices. Mobile phones and "dumb" devices (sensors and actuators) are being increasingly equipped with intelligence so that they are becoming able to act autonomously for supporting new and advanced applications for healthcare, transportation, business control, and security, as well as energy and environmental monitoring.

Several estimations have been made for the number of devices that will be interconnected in the next few years and it looks like that billions of devices will be connected to the global Internet by 2020. In such a hyperconnected world, where all these devices are continuously monitoring their environment including the activities and everyday lives of citizens new threats arise regarding security and privacy. Providing a holistic security framework for IoT systems is not an easy task to do, because it requires cross-layer mechanisms [1] and systems needs to be designed to be secure and privacy preserving. Retrofitting security mechanisms in non-secure IoT systems can provide only a very limited level of security.

The aim of this chapter is to discuss the challenges for security and privacy in a hyperconnected world where humans are assisted by machines and technology, but not watched by or through them. Starting from the need to adopt the essence of "security and privacy by design", we discuss why there is a need to embed security mechanisms in a system from the conceptual phase through the design process to ensure a maximum level of data protection and to guarantee end-to-end security.

Firstly, we put the focus on the devices discussing two different research areas: (i) physical IoT security, namely what are the threats to IoT when someone has access to the physical device and how can we protect them and (ii) embedded security and privacy on the constrained devices and why a system cannot be fully secure without securing the devices that generate the data first. This latter part discusses several techniques, e.g. for lightweight encryption, data minimization, integrity protection and usage of gateways to enforce security policies close to the constrained devices.

Secondly, this chapter discusses the importance of protecting not only the data, but the metadata as well to ensure that the communication stays unobservable, providing also countermeasures regarding how to be protected from network traffic analysis. Access control based on trust policies is also an important research area in the IoT and is briefly discussed next, aiming to show the importance of context information in the decisions regarding access control.

Finally, we conclude with a discussion on enforcing security and privacy in the "Cloud", as more and more IoT systems are utilizing the cloud both for storage and processing of the IoT data. What type of security and (mainly) privacy mechanisms need to be applied in the cloud to protect the data is currently an area which only lately started to receive attention and so far has not been properly explored, so we try to provide here an overview and suggest a way forward.

6.2 End-to-End Security and Privacy by Design

End-to-end security is a term that has quite distinct meanings depending on the OSI layer it refers to. In a hyperconnected IoT world, a multitude of networks and heterogeneous systems are bridged including a wide range of middleware systems that are all gathering, storing, and processing data. Then, from these huge amounts of data, information has to be generated to extract context for making smart decisions. Hence, end-to-end security between the devices and the applications is of paramount importance for protecting the privacy of people's personal data across the different systems and technologies that are involved. This requires strong data protection not limited to transit over wireless and intermediate Internet links, but also in all intermediate storage and processing points, till the data finally reaches solely its intended recipient.

The amount of acquired and processed data that will be ubiquitously provided in IoT becomes a huge concern for the people who are directly or indirectly monitored through their physical surroundings. Collection of personal information, starting even from their own devices and the surroundings they interact with, is high in quantity, quality and sensitivity. All this motivates the need for privacy in IoT [2].

The ubiquitous data collection in IoT is massive, even higher in comparison to other intrusive systems, such as online social networks and search engines [3]. While these generally trade privacy for commodity, their data collection depends on user interaction.

The ubiquity and pervasiveness of sensors to measure the status and context of an environment bring new types of privacy threats for the persons acting in that environment, regardless of them being users of the system or not. Thus, protecting the privacy of system participants as well as casual users and non-involved subjects in a future IoT is one of the main challenges for privacy-related research.

With the extensive data collection in mind it is clear that much of the business value lies in offering services that process and analyse the huge amounts of data collected [2, 4]. Nevertheless, these services should be as well privacy-enhanced, respecting and protecting the privacy of people's personal information. As a prerequisite for this the IoT systems must be built based on the concept of "privacy by design", which means that privacy enhancing mechanisms must be deeply rooted inside the IoT architecture. Furthermore, the solution should be such that every data subject should be able to give consent to the collection, storage and processing of their personal data for the particular known in advance purpose (consent and purpose). These are the challenges of "Privacy by Design" in IoT, see [5, 6].

Tackling these challenges is one of the most important business factors of the future IoT. As stated in the Opinion 8/2014 of the Article 29 Data Protection Working Party: "Organisations which place privacy and data protection at the forefront of product development will be well placed to ensure that their goods and services respect the principles of privacy by design and are equipped with the privacy friendly defaults expected by EU citizens."

6.3 Physical IoT Security

The major concern when implementing cryptographic functions on constrained devices is efficiency, due to the fact that devices are battery driven and shall be working for years. Unfortunately, this focus may lead to a vulnerable network even though cryptographic functions may be supported by those devices. The issue here is that implementations of cryptographic algorithms may be insecure even if the algorithm is considered to be secure. An implementation may indirectly provide information on the keys used for example by its timing or energy consumption. This is especially dangerous in the IoT context since here at least for some applications we need to consider that devices can be stolen, analysed in a well-equipped lab and brought back. Due to the wireless communication and the fact that the devices can be unreachable for some time such attacks might go fully undetected. Next, we provide examples of such attacks and their countermeasures.

6.3.1 Selected Low-Cost Attacks

The strength of cryptographic algorithms according to the definition of Kerckhoff [7] is based only on the used key that is kept secret. This means a potential attacker may know the algorithm itself, plain text, encrypted text and even the length of the key. From this point of view cryptographic approaches are secure if the time for brute forcing is long, that is if length of the key is sufficient.

The main assumption here is that the cryptographic device is a black box for an attacker, assuming he knows the cryptographic function but cannot get any details about how it is calculated. But in the IoT environments this assumption does no longer hold, due to the fact that devices may be stolen.

Even simple measurements like the ones of the current flowing through the chip or its electromagnetic radiation while a cryptographic function is calculated provide sufficient details to extract the key successfully. Such attacks – denoted as Side Channel Analysis (SCA) attacks – are often low-cost, easy and powerful. Even a single measurement can be sufficient to extract the cryptographic key in a few minutes for algorithms that are considered to be mathematically secure.

Figure 6.1 and Figure 6.2 show the same part of a measured PTs corresponding to processing the first 15 bits of the same cryptographic key using the same input on two different accelerators. It's a power trace of the elliptic curve point multiplication denoted as kP.

The calculations executed by two different IHP hardware accelerators of the kP operation of the standardized B-233 curve [8]. The shape of the measured traces is influenced by the private key, i.e. the shape of PTs while processing a '1' key bit differs from the shape of a '0'.

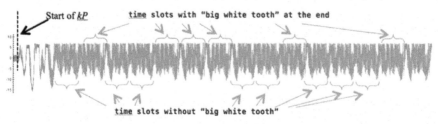

Figure 6.1 Implementation of cryptographic function – here the elliptic curve point multiplication – without paying attention to the SCA i.e. the shape of PT depends on the contents of the processed bit. This allows extracting the cryptographic key directly from the measured trace.

Figure 6.2 Implementation of the same cryptographic function taking SCA into account: the shape of the PT is always the same always, i.e. different key bits can no longer be distinguished in the PT. The cryptographic key cannot be extracted directly from measured trace.

If the cryptographic function is implemented without considering SCA this influence can be strong and the attacker can directly extract the key from a measured PT.

For example in Figure 6.1 two different kinds of the shapes are observable: time slots that have a "big white tooth" at its end and those without it. Using the assumption that the big white tooth at the end of the timeslots corresponds to the processing of a '1' key bit and other kind of slots corresponds to '0' key bit the used key can be correctly extracted.

Cryptographic algorithms implemented without considering SCA attacks can be called "weak" implementations. Knowledge about the details of such attacks – their main assumptions and the exploited characteristics – can help to implement the cryptographic algorithms in a way to be more resistant to such known attacks.

Figure shows the same part of the power trace of the same kP operation but executed on an improved version of our hardware accelerator, now all key bits are processed in the same way, i.e. simple power analysis attacks do not succeed.

Differential power analysis attacks are more powerful using statistical methods for analysis of measured traces. A very efficient, low-cost, fast and relatively easy attack on an elliptic curve cryptography (ECC) implementation is the horizontal power analysis using a difference of means test.

Each time slot can be observed as an independent curve. The mean curve of all slots can be calculated. After this the mean curve can be compared point wise with each slot.

If the power value of the mean curve is higher than the value of the current slot, it was assumed, this slot corresponds to the key bit value '1', otherwise to '0'. Thus, the first key candidate was obtained. Repeat this for all other points of the mean curve and you obtain the remaining key candidates.

We performed a horizontal power analysis attack using the difference of means test as described above for two simulated power traces. The power consumption of the IHP ECC design while processing the given EC point P using two different 232 bit long keys – k1 and k2 – was simulated using Synopsis Tools PrimeTime [9].

It obtained 57 key candidates for each of the investigated keys and we calculated the correctness of the extraction for each key. From a security point of view the ideal case is if the correctness of the key extraction is 50% for all key candidates. The green line in Figure 6.3 corresponds to this case.

Figure 6.3 demonstrates how powerful a difference of means based attack can be. In the case investigated here 225 bits of the 232 bit long 1st key

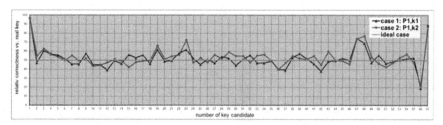

Figure 6.3 Relative correctness of the extraction of the key for each of the key candidates as a curve.

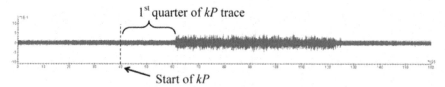

Figure 6.4 Difference of the traces of kP and of key candidate · P. The first quarter of the trace is equivalent to the noise, i.e. the quarter of the most significant bits of the key candidate is the same as the quarter of the most significant bits of k while all remaining bits differ.

candidate were extracted correctly, i.e. the correctness is about 97% in both keys. The correctness of the next probable 4 key candidates is also high from 70% up to 90%.

The next power analysis attack (Figure 6.4) that we performed based on direct comparison of two traces is similar to the one first introduced in [10]. The main assumption here is that an attacker can run the device with a key candidate.

The idea is that the difference of two power traces is close to zero, i.e. is comparable to the noise, if the kP operation with the same EC point and with the same scalar k is performed. This means the key can be extracted serially, bit by bit.

Using the attack sketched above only about 100 measurements without any statistical processing of the measured data are necessary to extract a 232 bit long key k correctly. So the mathematically strong secure 232 bits long cryptographic key can be extracted correctly in a few hours only.

6.3.2 Key Extraction Attacks and Countermeasures

Figure 6.5 represents all types of attacks and countermeasures for public key cryptography. The diagram reflects that most of the attacks and counter-measures are based on a never expressed assumption i.e. the fact that the

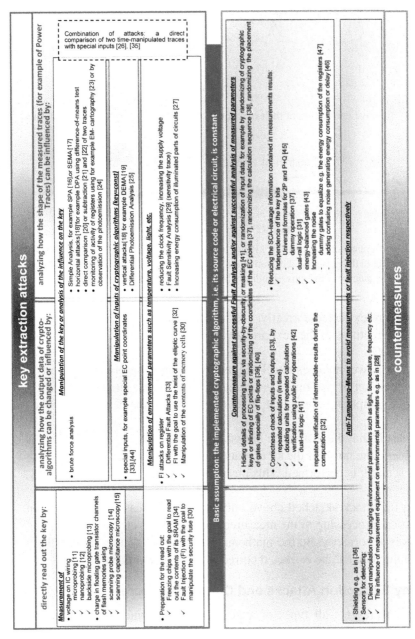

Figure 6.5   Key extraction attacks and countermeasures.

implementation of the cryptographic function is constant i.e. it cannot change during the attacks. This assumption is displayed as a rectangle connecting the attack and the countermeasure part.

The columns in the figure display:

- attacks in which the attacker can read out the key directly (left most column).
- changes in the output data (middle column).
- changes in the measured traces (right most column).

The rows represent all parameters that can be manipulated by the attacker i.e. the key candidate, other input data as well as the environmental parameter are given as rows.

The part of the Figure 6.5 representing the countermeasures is similarly structured as the attack part. The columns represent the same type of attacks as in the upper part of the diagram. The rows show countermeasures that (i) reduce the information that is contained in measurement results or that avoid access to faulty intermediate results, and (ii) avoid attacks by detecting the attack before it has any effect on the cryptographic implementation.

All countermeasures displayed in Figure 6.5 can help avoid attacks or at least hamper the potential success. Thus, they provide reasonable means to increase the security of the IoT. They come with some cost with respect to area and/or energy. Since cost of the devices and/or energy efficiency are paramount in IoT applications the use of countermeasures needs to be considered carefully. But in case physical access to IoT devices cannot be avoided and security is essential countermeasures need to be included in the implementations in order to ensure security.

6.4 On Device Security and Privacy

Designing a secure IoT system requires the embedding of security and privacy enhancing mechanisms locally on the devices near the physical entity of interest, whenever possible. Of course, this is much harder and more costly to maintain, e.g. it requires doing software updates for each smart device, which in turn requires reprogramming the actual device over the air, restarting it without needing human intervention and without configuring it again. Hardware must be capable of supporting advanced or even basic security mechanisms, as an insecure or non-private system design is hard to be turned later on into a privacy preserving or secure system.

The RERUM project tackles this with an "On-Device First" approach. RERUM's devices are made capable to run algorithms that enable the protection of security and privacy locally, by supporting advanced on-device security and privacy preserving mechanisms and over the air updating of the on-device software, while maintaining their energy consumption at very low levels.

6.4.1 Mediated Device Access for Security and Privacy

Security and privacy threats are continuously becoming more intelligent and they require more sophisticated countermeasures than IoT devices are capable of. Hence, we need an IoT gateway or IoT router to shield it. This is known as mediated device access. This gateway enables to hot-fix or firewall a large number of IoT devices from emerging threats, without the need to exchange every hardware device. Of course, if the local hardware device's privacy and security capabilities are outdated, the local threat level increases regardless of a gateway firewalling them from global threats. Thus, if one wants to secure the hotel building's management from the attacking hotel guest, each local device's security must be kept up-to-date.

Additionally to security, a gateway could be the local point of control and enforcement for privacy, as it has far more processing capabilities and gathers far more information from the environment than a single device. We assume that to apply privacy enhancing technologies (PET) the gateway would be trusted to act in the data subject's interest. Moreover, the gateway can use the diverse information it has from fusing other data from the data subject's devices as some form of ground truth or guidance, e.g., apply the PET differently when the data subject is at home or not. Mediated access to the lower end IoT devices, and hence some IoT gateway, is a necessity to ensure security and privacy.

6.4.2 Encryption

IoT mainly consists of severely resource constrained devices that are not capable of running complex encryption mechanisms like standard PCs. Thus, lightweight encryption mechanisms are of paramount importance for increasing the security of IoT. Lightweight cryptography normally provides adequate security but does not always consider energy efficiency. Symmetric key cryptography using Advanced Encryption Standard (AES) [48] is widely used in practical implementation of encryption based on block ciphers on constrained devices. Hash functions (e.g. SHA-3 [49]) are also widely used but they are not lightweight, and only lately there are some research steps towards lightweight hash functions. Elliptic Curve Cryptography (ECC) [50] is used in IoT due to the fact that it uses keys of much smaller size than standard public key cryptography mechanisms. However, its execution time might still not be fast enough for some devices.

The majority of existing encryption algorithms do not fully fulfil the requirements for energy efficiency. Furthermore, key distribution schemes are necessary for their proper operation, making the network vulnerable to adversaries that manage to capture the keys during key exchange. Basic requirements for efficient lightweight IoT encryption can be assumed to be the following:

- Encryption mechanisms have to be optimized for their energy efficiency. This is critical as sensors are resource constrained devices in terms of memory, CPU, and processing;
- Key distribution schemes should be avoided or their usage should be minimized. These consume valuable energy, and there is also the risk of information hijacking (by an adversary) during the key exchange and
- Keys should not be pre-stored on the sensor device (currently this is done usually during manufacturing). This poses a significant security threat as sensors can be easily compromised when placed in outdoor environments.

In Wireless Sensor Networks (WSNs) the Compressive Sensing (CS) technique has been widely used for compressing the data that are gathered by sensors. CS is a very useful technique because it applies at the same step both data compression and lightweight lossy encryption [51]. The reconstruction error is directly related with the level of compression and encryption and the nature of the signal that is captured by the sensor. For example, a slow varying temperature signal has very low reconstruction error, while another signal that has rapid changes will result to a very high reconstruction error.

Within RERUM, a technique for extracting the encryption keys for CS at real-time has been proposed, supporting the requirement for not hardware-coding the keys on the IoT devices [52]. Key extraction is performed using channel measurements, thus there is no need for any key distribution mechanism. The derived keys are used for encryption/decryption using the primitives of CS. Evaluation results have shown that legitimate nodes experience a very low reconstruction (decryption) error, while adversaries located at a distance greater than half of the carrier frequency's wavelength, experience a higher error, thus being unable to capture and decode sensitive information.

6.4.3 Integrity

Integrity is the "property that data has not been altered [...] in an unauthorised manner"[1]. In a hyperconnected world, the IoT's flow of communication is highly loosely coupled, meaning that data that are transmitted over a secure channel are then stored and processed or transmitted further later. Protecting the integrity for those type of loosely connected data can be achieved by message-level protection mechanisms. Using a cryptographically secure signature scheme, based on asymmetric keys, allows verifying that data has not been modified in unauthorised ways. Additionally, you gain origin-authentication, i.e., verifying which entities' public key signed the data. Adding a message authentication code (MAC), with a shared key between sender and receiver, also allows ensuring that the message's integrity cannot be violated without being detected by the receiver.

6.4.4 Data Minimisation

In [53], the authors underline that the very foundation of privacy by design is data minimization, which is defined as the property to limit as much as possible the release of personal data and, for those released, preserve as much unlinkability as possible [54]. To exemplify how data minimization is related to privacy by design, the reader is referred to the popular Privacy By Design framework [55].

If personal data collection is minimized from the very beginning, much less effort will be needed to further define and implement appropriate privacy enhancing mechanisms. The application of adequate technologies for data minimization requires expertise in the services that the IoT system provides. The engineer must decide if it is possible to render the same (or comparable) functionality with less amount of personal information. In some cases, unlinkability might not always be desirable, for instance if devices and data must be needed to be linked to a user, for billing, authentication or otherwise.

The best place to achieve data minimization is on the devices where the data are sensed, as the amount of personal information can be minimized before the data are transmitted from the devices to the backbone system. This can be enforced with hard privacy mechanisms, such as malleable signatures and group signatures [56], which can be implemented on devices to ensure integrity and create unlinkability for data. Location privacy technologies [57] can be applied on devices e.g. to measure traffic data and compute averages

---

[1]ETSI TS 133 105 V10.0.0 (2011-04)

of speed and distance, while anonymizing a participant's real location in the geolocation system.

In addition to privacy preserving technologies for sensed data, further privacy mechanisms are needed for quasi-identifiers such as metadata and IP-addresses can provide sensitive information. Traffic analysis, as one example, is frequently used to identify the sources of data and thus de-anonymize the information. Mechanisms to ensure communication observability can further enhance privacy protection for the IoT, which are discussed in the following section.

6.5 Unobservable Communication

Even if the protection of user data is addressed by means of end-to-end encryption in the future, we still need to look into information loss caused by leaking protocol metadata. This leakage can go up to the point, which may render end-to-end encryption obsolete. To reduce it, at least the following properties [58] shall be preserved by the network of IoT devices:

- Coding – All messages with the same encoding can be traced.
- Size – Messages with the same size can be correlated.
- Timing – By observing the duration of a communication and considering average round-trip times between the communication partners patterns of network participation can be extracted.
- Counting – The number of messages exchanged between the communicating parties can be observed.
- Volume – Volume combines information gained from message size and count. The volume of data transmitted can be observed.
- Pattern – By observing communication activity, patterns of sending and receiving can be observed.

Furthermore, message frequencies and flow can be analysed. The message flow between parties includes both the traffic volume and communication pattern. Communication partners have a unique distinguished behaviour that can be fingerprinted. An observer can perform a brute force analysis of the network by observing all possible paths of communication and generating a list of all possible recipients.

Finally the observer can also perform a long term intersection/disclosure analysis of the network by observing devices and the network for long time and reducing the set of possible communication paths and recipients by analysing online and offline periods. Characteristic usage patterns, such as an IoT device connecting every minute, may appear and can be used to further reduce the number of possible paths.

The following Table summarises the message properties and how they can be addressed.

Table 6.1 Message properties

Attacks Based on	Proposed Solutions
Message Coding	Change coding during transmission e.g. with k-nested encryption
Message Timing	1) batched forwarding of messages 2) random delay of messages ($delay_{min} \geq latency_{max}$)
Message Size	Use a predefined message size and padding small messages
Message Counting	Receive and forward a standard number of messages and use dummy traffic
Communication Volume	Protect message size and communication volume
Communication Pattern	Continuous network participation
Message Frequency	Use a standardized message exchange pattern
Brute Force	No clear protection dummy traffic helps
Long Term Intersection	No clear protection continuous connectivity and dummy traffic help

6.5.1 Resisting Network Traffic Analysis

Leakage of metadata can be reduced by providing protection against network traffic analysis. This includes endpoints, timing and location information. Traffic analysis can be addressed by ensuring unobservable communication as implemented by anonymising networks using generally proxy chains. Anonymising proxy networks have started with the implementation of Chaum's Mix in 1981 [59]. The system tunnels encrypted traffic through a number of low-latency proxies, as depicted in Figure 6.6.

Initially, interest in this field was primarily theoretical but in the last 30 years a lot of research in this field has looked at developing practical and usable systems for preserving anonymity [60, 61]. Such systems cover Email, Web browsing and other services like peer-to-peer networks and IRC chat. Systems like The Onion Router (TOR) and the Invisible Internet Project (I2P) allow generic layer 3 transmission. While TOR was primarily developed to

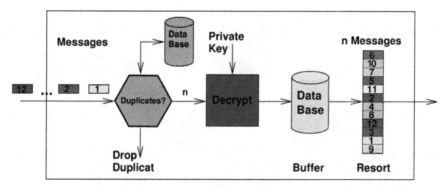

Figure 6.6 Chaum's MIX.

allow anonymous web browsing in close to real-time the general concept is applicable to prevent traffic analysis in the IoT network.

Once traffic leaves the TOR network it can be observed, therefore end-to-end encryption is needed and is the responsibility of the end nodes. Apart from TOR, there is I2P, an anonymous/pseudonymous network layer. Like TOR, I2P can be extended for many services. I2P is neither as secure nor as fast as TOR, but can handle large volumes of traffic, like those foreseen for the IoT.

6.6 Access Control Based on Policy Management

A policy management framework (as developed in the iCORE project) supports IoT-specific access control requirements like the hyperconnected-ness and distributed-ness of the IoT and the need of applications to share resources and even data. The Security Toolkit (SecKit) [62] models the IoT system for security specification purposes. The system design is divided into an entity domain and a behaviour domain, with an assignment relationship between entities and behaviours.

In the entity domain, the entities and the communication mechanisms allowing the entities to exchange information are specified. In the behaviour domain the behaviour of each entity is detailed including actions, interactions, causality relations, and information attributes. It is also possible to specify the data, identity, context, trust, role, risk, and security rules in so called metamodels. Figure 6.7 illustrates their dependencies.

The context metamodel specifies Context Information and Context Situation types. Context Information is a simple type of information about an entity that is acquired at a particular moment in time, and Context Situations are a

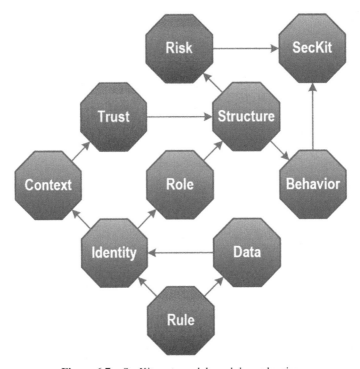

Figure 6.7   SecKit metamodels and dependencies.

complex type that models a specific condition that begins and finishes at specific moments in time [63]. For example, the "Body Temperature" is a Context Information type, while "Fever" is a situation where a target patient has a temperature above 37 degrees Celsius. Entities are associated to context situations using roles (e.g. patient).

A Context Manager component monitors and registers events when situations begin and end. These events contain references to the entities that participate in the situation and can be used to support the specification of the policy rules. Policy rules can be specified to represent authorizations to be granted when a situation begins and data protection obligations that should be fulfilled when the situation ends. For example, access to the patient data can be allowed when an emergency situation starts with the obligation that all data is deleted when the emergency ends. A security policy may be specified to allow access to data when the situation starts and to trigger the deletion of the data when the situation ends.

The security policies have to be disseminated to the device that is gathering the data under consideration in a secure way. Depending on the security policy, the device has to trigger and apply the appropriate mechanism for transmitting the data in the exact format needed by the application. This includes a two-step process; (i) at first the device has to map the policies for the application to specific data gathering policies and (ii) then it should identify the encryption/security level of the data to identify the proper transmission mechanisms, considering also the energy efficiency requirements of the devices (using i.e. an adaptive encryption scheme). For example, in a traffic monitoring scenario, users in cars may be sending information regarding traffic to an application, which should know only how much traffic there is at every street segment. The users' phone has the ability to send various types of traffic related data, i.e. exact location every second, speed every second, direction of movement, etc. If the application wants to estimate the traffic, the related policies should be considered by the devices of the users, so only an average speed per time period and street segment is sent, in order to avoid disclosing the exact location of the user at each point of time (ensuring privacy by design). Actually, intermediate nodes (i.e. the gateway) should also consider these policies and send to the application server only aggregated/average data so that the location of the users will be hidden from the application point of view. Other applications that need to know the exact location of the user (depending on their access control policies) will indeed be identified as such by the devices, which will transmit the exact location (i.e. for a person to track his car if it is stolen). It is evident, thus, that the transmission of the security policies to the devices is of crucial importance for ensuring the security and privacy of the overall system. The system should be able to identify the integrity of the policies that are sent to the devices, so that unauthorized applications will not gain access to privacy-sensitive data.

The security rules model supports the specification of rule templates (a.k.a. policies) to be enforced and the configuration rules to instantiate these templates. Templates can be specified considering the security and privacy non-functional requirements of confidentiality, data protection, integrity, authorization, and non-repudiation. The security rule templates are Event-Condition-Action rules, with the Action part being an enforcement action of Allowing, Denying, Modifying, or Delaying an activity carried out by an IoT device or application. Furthermore, the Action part may also trigger the execution of additional actions to be enforced, or to specify trust management policies to increase/decrease the trust evidence for a specific trust aspect.

From a trust management perspective, the SecKit supports the specification of aspect-specific trust relationships and exchange of trust recommendations. For example, trust relationships can be defined for identity provisioning aspect, privacy protection, data provisioning, and so on. A trust relationship also includes a trust degree, which is mapped to a Subjective Logic (SL) opinion considering the amount of belief, disbelief, and uncertainty [64]. The aspect-specific approach considering uncertainty is more realistic from a human perspective since people usually trust others for specific purposes (e.g., a mechanic to fix your car) and most of the time cannot be absolutely certain about the amount of trust they may place.

The security policy rules can be delegated from one administrative domain to another when the domains interact and exchange data. For example, when a smart home exchanges data with a smart vehicle, the smart home can exchange the policies that regulate the authorizations and obligations associated to the exchanged data that should be enforced by the smart vehicle. This delegation of sticky flow policies must be supported by trust management mechanisms [63] in order to guarantee or increase the level of assurance with respect to the enforcement of the policy rules by the smart vehicle.

6.7 Security and Privacy in the IoT Cloud

The "Cloud" complements quite well the IoT supporting the storage and processing of the large amounts of data that are gathered by constrained devices. However, the Cloud introduces new threats for security, but especially with respect to information privacy. When IoT data are moved to the cloud for storage we could use encryption to protect it. However, if the application turning that data into information is running in the cloud, then the cloud provider becomes yet another third party that needs to process the stored data gathered from the physical world. Hence, the provider inherits all the privacy problems of the data. In fact, the third party becomes part of the IoT application provider's own computation and storage infrastructure. However, the cloud provider is technically not under its full control. This situation has shown to be problematic and incidents recently showed that economic incentives and legal tools used to increase trust in the service provider, e.g. Service Level Agreements, are by far not sufficient to guard personal data and trade secrets against illegal interceptions, insider threats, or vulnerabilities exposing data in the cloud to unauthorized parties. While being processed by a cloud provider, data are typically neither adequately protected against unauthorized read access, nor against unwanted modification, or loss of authenticity. Consequently, in the most prominent cloud deployment model today – the public cloud – the cloud service provider (CSP) necessarily needs to be trusted. Next, we will provide some selected areas from PRISMACLOUD's cryptographic research and highlight their foreseen suitability for IoT data.

6.7.1 Verifiable and Authenticity Preserving Data Processing

Verifiable computing allows checking the result of a computation for its validity, even if the computation itself was done by one or more untrusted processing units. While respective systems have already been implemented, they have not yet seen real-world deployment. Besides general purpose systems [65] there are various approaches that are optimized for specific (limited) classes of computations or particular settings [66]. A cloud user can facilitate those mechanisms to check if collected measurements have been processed correctly, and, if not so, they can identify (maliciously) incorrect calculations.

When data are subject to computations executed by the cloud provider, it is extremely helpful if the processing allows preserving the authenticity of data that are manipulated by computations. The most generic tool for preserving authenticity under admissible modifications are (fully) homomorphic signatures (or message authentication codes) [67]. Signatures with more restricted capabilities, like redactable signatures introduced in [68, 69], offer a restricted set of capabilities, but with better performance [70, 71]. Redactable and sanitizable signatures have been proven to strongly preserve the privacy [72] of the original values when they have been updated/changed or redacted [73]. They allow preserving the authenticity on data introduced by an IoT device's signature, which vouches for the data's origin, even after processing. Thus, the cloud user after authorized processing can still verify the involved data's authenticity.

6.7.2 Structural Integrity and Certification of Virtualized Infrastructure

Structural integrity and certification of virtualized infrastructures connects attestation of component integrity, i.e., proving the trustworthiness of claims about the infrastructure, and security assurance of cloud topologies, i.e., guaranteeing that a cloud topology provides certain security guarantees (e.g., network isolation). This is a clear benefit for cloud infrastructure consumers as their confidence in infrastructure properties can be increased and the cloud provider can be held accountable. The recent concept of graph signatures [74] is a promising candidate to connect the two aforementioned areas. They allow a trusted third-party auditor to digitally sign a graph and prove in zero-knowledge properties of the graph, such as connectivity or isolation. Graph signatures can be a handy tool for a cloud provider to convince a certain customer in a multi-tenant environment that the granted infrastructure fulfils certain security properties, but at the same time to not disclose the confidential blueprint of the virtualized infrastructure. For instance, the cloud may prove that a customer's part of the infrastructure is isolated from other tenants without revealing how their part of the infrastructure looks like. A first implementation of a system that allows certification of and proofs about a certified infrastructure as well as other relevant and interesting use-cases has already been outlined in [75]. This allows to attest an IoT's infrastructure in a way to ensure that certain security properties are satisfied and improves accountability.

6.7.3 Privacy Preserving Service Usage and Data Handling

Privacy-preserving service usage essentially means to realize 1) data minimisation, i.e., to only reveal information that is essential for service delivery, and 2) avoid (behavioural) tracking of service users. This is especially important in cloud based applications, as such information may, among others, reveal confidential business information [76]. Attribute-based anonymous credential (ABC) systems and related concepts such as group signature schemes [77] are important concepts for realizing such privacy-preserving applications. They allow users to authenticate in an anonymous way, i.e., without revealing their identity, but allow to prove claims that enable a service provider to still make access decisions. Although they are quite mature in the research community, they still lack practical adoption, which, however, needs to be considered as a very important topic for future cloud IoT applications.

Another issue is privacy in context of data handling. In contrast to achieving data privacy by means of encryption, which realizes an all-or-nothing mechanisms for the access to the data, we thereby mean scenarios which are often encounter when processing of data by third parties in the cloud is required. Essentially, this covers mechanisms for data anonymisation such that a provable level of anonymity can be achieved, i.e., k-anonymity [78] or differential privacy [79]. In particular, one requires a guarantee that when (large amounts of) structured data are given away or are dynamically queried, it can be ensured that a targeted degree of privacy is guaranteed, i.e., data collected from many individuals does not allow to uniquely identify single individuals but still allows to compute meaningful statistical parameters. Techniques for privacy-preserving service usage allow IoT devices to anonymously authenticate to services and prevent linking of transactions conducted by IoT devices. Data anonymisation can help to protect privacy of individuals if IoT devices send sensitive information (e.g., health data) to the cloud and the data is later released for further processing.

6.7.4 Confidentiality of (Un-)structured Data

Confidentiality of data when outsourced to the cloud for the purpose of storage and/or processing is considered to be sine qua non, since cloud providers can neither be considered as fully trustworthy nor are resistant to attacks. Encryption is a classical tool to provide confidentiality. Unfortunately, encryption clearly limits the functionality (how to operate on data), adding encryption to legacy applications may cause serious problems and the management of the involved cryptographic keys soon becomes highly complex. Within the last years, significant research has been put into cloud storage solutions that distribute the data to multiple clouds (aka cloud-of-cloud approach) [80]. They allow providing confidentiality for data at rest with strong security in a key-less manner under some non-collusion assumption and thus solve the key management problem (at least partially). An interesting challenge is to design such a distributed architecture using active nodes to fully delegate secure multi-user storage to the cloud. Thereby, the use of efficient Byzantine protocols helps to improve robustness and various types of secret sharing protocols can help to cope with different adversary settings. Furthermore, for a multi-user setting a trustworthy distributed access control mechanism is required and it is interesting to extend it with access privacy features. Another issue, as mentioned above, is the integration of encryption into legacy (e.g., database) applications, as they may be unable to use or store encrypted data, causing them to crash or alternatively, to output incorrect values. Techniques like format-preserving encryption (FPE) [81], order-preserving encryption (OPE) [82] and tokenization schemes have emerged as very useful tools as they can be directly applied without adapting the application itself.

6.7.5 Long Term Security and Everlasting Privacy

Classical cryptographic primitives such as digital signature schemes and encryption schemes are valuable tools to achieve integrity, authenticity, and confidentiality. If these properties, however, need to hold in the long-term, e.g., for some decades or even indefinitely, these tools often fail. Cryptanalytic progress and advances in computing power can reduce their security or may even make them entirely worthless. There are only few approaches that consider long-term confidentiality, integrity and authenticity. Moreover, many of the existing solutions lack in providing these properties [83, 84].

6.7.6 Conclusion

At the moment privacy guarantees with respect to user's IoT gathered data in the cloud can only be given on a contractual basis and rest to a considerable extent on organizational (besides technical) precautions. Companies or individuals alike are in the end a cloud user, and they themselves are responsible for their data's privacy, whether processing gets outsourced to the cloud or not.

Therefore, the H2020 project PRISMACLOUD is looking into novel security and privacy preserving methods, such that cloud usage can be facilitated even by organizations that deal with highly sensitive data such as health data and maintaining security critical services. PRISMACLOUD only just started in the first quarter of 2015. But the vision is that only a new generation of cryptographically secured cloud services with security and privacy built in by design can lead the way to achieving the required privacy properties for outsourced data storage and processing at the upper end of the IoT – privacy in the cloud.

6.8 Outlook

Security and privacy in the IoT world are research areas that only lately have attracted the attention of both the research and the industrial world. Up until now, the focus was limited on creating efficient middleware platforms to enable the services to gather data from the devices. This resulted in existing IoT deployments that are not secure and gather all types of personal information. Fortunately, recently the significant focus on security and privacy has resulted in important achievements not only in the technology domain, but also on the way the world sees the IoT. Security and privacy are now basically seen as the key points for the wider adoption of the IoT applications by the general public. If the citizens can be reassured that the IoT will not harm them, will not steal their private information and will not affect their lives in a negative way, only then they will gladly accept and embrace IoT and the full potential of IoT can unfold to improve their – and everyone's – quality of life.

This chapter presented a cross-layer approach on improving the security and the privacy of IoT systems, allowing them to work for the benefit of the people, without leaking information, presenting a risk or damaging people's privacy. Designing a system so complex as the IoT whilst guaranteeing that a certain level of security is achieved is an extremely complex and tedious task, and can not be retrofitted, so we must already design the IoT with privacy and security in mind. It is widely acknowledged that the security of an IoT system depends heavily on the devices, so we need to physically secure the IoT devices, as every system's level of security is as good/high as the one of its weakest part.

Although encryption can really contribute to protecting the data that are being exchanged in an IoT system, this is not quite enough. Even with encryption deployed end-to-end, the IoT still leaks information by communication metadata. If the volume and quality of the information collected is sufficiently large, even encrypted information can be extracted without breaking the encryption of the communication channel.

From what was previously described, it is also quite important to design the system to be privacy preserving, starting from embedding in the devices mechanisms for both data minimization and for enhancing privacy. These are quite important to ensure that the services will only get the exact data they need and nothing more, to avoid the possibility of linking data.

But we need to think even broader, the problem of privacy – and of security – well extends into the cloud. The society has to be able to trust the whole IoT value chain all the way up to the cloud. Thus, new cryptographically proven security and privacy mechanisms must be developed to allow provably using cloud services securely and privately.

In general, there is a lot of work done in the IoT world towards enhancing the security and the privacy of IoT systems. However, making significant progress in this area through research is not enough. The industrial world and the businesses need to put more focus on embracing and adopting security and privacy solutions. To complete the picture, regulations for protecting IoT data need to be put into place, to ensure the adherence of every player to the socially accepted norms of privacy in the EU. Only then the hyperconnected world of the IoT becomes not a threat to the citizens, but a useful tool to improve not only our's – but everyone's – everyday lives.

Acknowledgment

This work is partially funded by the EU FP7 projects RERUM (GA no 609094), SMARTIE (GA no 609062), iCore (GA no 287708) and the H2020 project PRISMACLOUD (GA no 644962).

Bibliography

[1] Henrich C. Pohls, et al. RERUM: Building a reliable IoT upon privacy- and security-enabled smart objects. *Wireless Communications and Networking Conference Workshops (WCNCW)*, IEEE, 2014.

[2] Jacob Kohnstamm, Drudeisha Madhub: Mauritius Declaration on the Internet of Things. *In: 36th International Conference of Data Protection and Privacy Comissioners*, 2014.

[3] Ralph Gross, and Alessandro Acquisti. Information revelation and privacy in online social networks. *Proceedings of the 2005 ACM workshop on Privacy in the electronic society. ACM*, 2005.

[4] Elgar Fleisch: What is the Internet of Things? An Economic Perspective. *In: Economics, Management, and Financial Markets 2* (2010), S. 125–157.

[5] Rodrigo Roman, Jianying Zhou, Javier Lopez: On the features and challenges of security and privacy in distributed internet of things. *In: Computer Networks 57 Nr. 10, 2266–2279. Towards a Science of Cyber Security Security and Identity Architecture for the Future Internet*, 2013.

[6] Marc Langheinrich: Privacy by Design — Principles of Privacy- Aware Ubiquitous Systems. Version: 2001. http://dx.doi.org/10.1007/3- 540-45427-6_23. In: ABOWD, GregoryD. (Hrsg.); BRUMITT, Barry (Hrsg.); SHAFER, Steven (Hrsg.): Ubicomp 2001: *Ubiquitous Computing Bd.* 2201. Springer Berlin Heidelberg, 2001. – ISBN 978-3-540- 42614-1, 273–291.

[7] D. Hankerson, A. Menezes, S. Vanstone: *Guide to Elliptic Curve Cryptography*. Springer-Verlag New York, Inc. (2004).

[8] National Institute of Standards and Technology: Digital Signature Standard (DSS), FIPS PUB 186-4, July 2013, http://nvlpubs.nist.gov/ nistpubs/FIPS/NIST.FIPS.186-4.pdf

[9] Synopsis – PrimeTime, http://www.synopsys.com/Tools/Implementat ion/SignOff/Pages/PrimeTime.aspx

[10] T. S. Messerges, E. A. Dabbish, R. H. Sloan: Power Analysis Attacks of Modular Exponentiation in Smartcards. *Proceedings of the Workshop on Cryptographic Hardware and Embedded Systems* – CHES 1999, LNCS Volume 1717, pp. 144–157, Springer Berlin Heidelberg, 1999.

[11] Z. Dyka, P. Langendörfer: Improving the Security of Wireless Sensor Networks by Protecting the Sensor Nodes against Side Channel Attacks. *Wireless Networks and Security, Signals and Communication Technology* 2013, pp. 303–328, Springer Berlin Heidelberg, 2013.

[12] Hitachi: Nano-Probing System, http://www.hitachi-hitec.com/global/em/nan/nan_index.html

[13] C. Helfmeier, D. Nedospasov, C. Tarnovsky, J. S. Krissler, C. Boit, J.-P. Seifert: Breaking and Entering through the Silicon. *Proceedings of the 2013 ACM SIGSAC Conference on Computer & Communications Security* – CCS 2013, November 4–8, 2013, Berlin, Germany, pp. 733–744.

[14] Ch. De Nardi, R. Desplats, Ph. Perdu, F. Beaudoin, J. L. Gauffier: EEPROM Failure Analysis Methodology: Can Programmed Charges Be Measured Directly by Electrical Techniques of Scanning Probe Microscopy? *Proceedings of the 31st International Symposium for Testing and Failure Analysis* – ISTFA 2005, November 6–10, 2005, San Jose, CA, USA.

[15] Ch. De Nardi, R. Desplats, Ph. Perdu, Ch. Guérin, J. L. Gauffier, Th. B. Amundsen: Direct Measurements of Charge in Floating Gate Transistor Channels of Flash Memories Using Scanning Capacitance Microscopy. *Proceedings of the 32nd International Symposium for Testing and Failure Analysis* – ISTFA 2006, November 12–16, 2006, Austin, Texas, USA.

[16] S. A. Kadir, A. Sasongko: Simple power analysis attack against elliptic curve cryptography processor on FPGA implementation, *Proceedings of International Conference on Electrical Engineering and Informatics*, pp. 1–4 , 17–19 July 2011, IEEE.

[17] E. De Mulder, P. Buysschaert, S. B. Ors, P. Delmotte, B. Preneel, G. Vandenbosch, I. Verbauwhede: Electromagnetic Analysis Attack on an FPGA Implementation of an Elliptic Curve Cryptosystem. *EUROCON 2005—International Conference on Computer as a Tool,* November 21–24, 2005, Belgrade, Serbia and Montenegro, pp. 1879–1882.

[18] C. Clavier, B. Feix, G. Gagnerot, M. Roussellet, V. Verneuil: Horizontal Correlation Analysis on Exponentiation. *Proceedings of the 12th International Conference on Information and Communications Security* – *ICICS 2010*, December 15–17, 2010, Barcelona, Spain, LNCS Volume 6476, pp. 46–61, Springer Berlin Heidelberg, 2010.

[19] E. De Mulder, S. B. Ors, B. Preneel, I. Verbauwhede: Differential Electromagnetic Attack on an FPGA Implementation of Elliptic Curve Cryptosystems. *World Automation Congress – WAC*, July 24–26, 2006, Budapest, Hungary.

[20] M. Hutter, M. Kirschbaum, T. Plos, J. M. Schmidt, S. Mangard: Exploiting the Difference of Side-Channel Leakages. *Proceedings of the Third International Workshop on Constructive Side-Channel Analysis and Secure Design – COSADE 2012*, May 3–4, 2012, Darmstadt, Germany, LNCS Volume 7275, pp. 1–16, Springer Berlin Heidelberg, 2012.

[21] Z. Dyka, Th. Basmer, Ch. Wittke, P. Langendoerfer: *Individualizing Electrical Circuits of Cryptographic Devices as a Means to Hinder Tampering Attacks*, Cryptology ePrint Archive 2015/442.

[22] T. S. Messerges, E. A. Dabbish, R. H. Sloan: Power Analysis Attacks of Modular Exponentiation in Smartcards. *Proceedings of the Workshop on Cryptographic Hardware and Embedded Systems – CHES 1999*, LNCS Volume 1717, pp. 144–157, Springer Berlin Heidelberg, 1999.

[23] J. Heyszl, S. Mangard, B. Heinz, F. Stumpf, G. Sigl: Localized Electromagnetic Analysis of Cryptographic Implementations. *Proceedings of the Cryptographers' Track at the RSA Conference – CT-RSA 2012*, San Francisco, CA, USA, February 27–March 2, 2012, LNCS Volume 7178, pp. 231–244, Springer Berlin Heidelberg, 2012.

[24] A. Schlösser, D. Nedospasov, J. Krämer, S. Orlic, J. P. Seifert: Simple Photonic Emission Analysis of AES. *Proceeding of Cryptographic Hardware and Embedded Systems – CHES 2012*, LNCS Volume 7428, pp. 41–57, Springer Berlin Heidelberg, 2012.

[25] J. Krämer, D. Nedospasov, A. Schlösser, J.-P. Seifert: Differential Photonic Emission Analysis. *Constructive Side-Channel Analysis and Secure Design – COSADE 2013*, LNCS Volume 7864, 2013, pp. 1–16, Springer Berlin Heidelberg, 2013.

[26] P.-A. Fouque, F. Valette: The Doubling Attack – Why Upwards Is Better than Downwards. *Cryptographic Hardware and Embedded Systems – CHES 2003*, LNCS Volume 2779, pp. 269–280, Springer Berlin Heidelberg, 2003.

[27] S. Skorobogatov: Optically Enhanced Position-Locked Power Analysis. Proceedings of 8th International Workshop *Cryptographic Hardware and Embedded Systems – CHES 2006*, Yokohama, Japan, October 10–13, 2006, LNCS Volume 4249, Springer Berlin Heidelberg, 2006.

[28] Cl. Helfmeier, Chr. Boit, U. Kerst: On Charge Sensors for FIB Attack Detection. *Proceedings of the IEEE International Symposium on Hardware-Oriented Security and Trust – HOST 2012*, San Francisco, CA, USA, Jun. 2012, pp. 128–133.

[29] H. Sakamoto, Y. Li, K. Ohta, K. Sakiyama: Fault Sensitivity Analysis against Elliptic Curve Cryptosystem. *Proceedings of the 2011 Workshop*

on Fault Diagnosis and Tolerance in Cryptography – FDTC 2011, pp. 11–20.

[30] S. P. Skorobogatov: *Semi-invasive attacks – a new approach to hardware security analysis.* Technical Report UCAM-CL-TR-630, University of Cambridge, Computer Laboratory, April 2005.

[31] S. Moore, R. Anderson, P. Cunningham, R. Mullins, G. Taylor: Improving Smart Card Security using Self-timed Circuits, *Proceedings of the International Symposium on Advanced Research in Asynchronous Circuits and Systems* – ASYNC 2002, pp. 211–218.

[32] P. Fouque, R. Lercier, D. Real, F. Valette: Fault Attack on Elliptic Curve with Montgomery Ladder Implementation. *Proceedings of the 5th Workshop on Fault Diagnosis and Tolerance in Cryptography – FDTC 2008*, August 10, 2008, Washington, DC, USA, pp. 92–98.

[33] I. Biehl, B. Meyer, V. Müller: Differential Fault Attacks on Elliptic Curve Cryptosystems. *Proceedings of the 20th Annual International Cryptology Conference – CRYPTO 2000*, Santa Barbara, CA, USA, August 20–24, 2000, pp. 131–146, Springer Berlin Heidelberg, 2000.

[34] S. Skorobogatov: *Low temperature data remanence in static RAM.* Technical Report UCAM-CL-TR-536, University of Cambridge, Computer Laboratory, June 2002.

[35] N. Homma, Atsushi Miyamoto, Takafumi Aoki, Akashi Satoh, Adi Shamir: Collision-based Power Analysis of Modular Exponentiation Using Chosen-message. *Proceedings of the 10th International Workshop – CHES 2008*, LNCS Volume 5154, pp. 15–29, Springer Berlin Heidelberg, 2008.

[36] S. Briais, J.-M. Cioranesco, J.-L. Danger, S. Guilley, D. Nacchache, Th. Porteboeuf: Random Active Shield. *Proceedings of the Workshop on Fault Diagnosis and Tolerance in Cryptography – FDTC 2012*, September 9, 2012, Leuven, Belgium, pp. 103–113.

[37] J. Coron: Resistance against Differential Power Analysis for Elliptic Curve Cryptosystems. *Proceedings of the First International Workshop – CHES 1999*, August 12–13, 1999, Worcester, MA, USA, LNCS Volume 1717, pp. 292–302, Springer Berlin Heidelberg, 1999.

[38] E. Oswald, M. Aigner: Randomized Addition-Subtraction Chains as a Countermeasure against Power Attacks. *Proceedings of the Third International Workshop – CHES 2001*, May 14–16, 2001, Paris, France, LNCS Volume 2162, pp. 39–50, Springer Berlin Heidelberg, 2001.

[39] K. Itoh, T. Izu, and M. Takenaka: A Practical Countermeasure against Address-Bit Differential Power Analysis. *Cryptographic Hardware and*

Embedded Systems – CHES 2003, Proceedings of the 5th International Workshop, Cologne, Germany, September 8–10, 2003, LNCS Volume 2779, pp. 382–396, Springer Berlin Heidelberg, 2003.

[40] J. Heyszl: *Impact of Localized Electromagnetic Field Measurements on Implementations of Asymmetric Cryptography*. Technische Universität München, Lehrstuel fuer Sicherheit in der Informationstechnik an der Fakultaet fuer Elektrotechnik und Informationstechnik, Dissertation, 2013.

[41] J.-S. Coron, L. Goubin: On Boolean and Arithmetic Masking against Differential Power Analysis. *Proceedings of the Second International Workshop – CHES 2000*, Worcester, MA, USA, August 17–18, 2000, LNCS Volume 1965, pp. 231–237, Springer Berlin Heidelberg, 2000.

[42] A. Shamir US 5991415.

[43] K. Tiri, M. Akmal, I. Verbauwhede: A Dynamic and Differential CMOS Logic with Signal Independent Power Consumption to Withstand Differential Power Analysis on Smart Cards. *Proceedings of the 28th European Solid-State Circuits Conference – ESSCIRC 2002*, September 24–26, 2002, pp. 403–406.

[44] L. Goubin: A Refined Power-Analysis Attack on Elliptic Curve Cryptosystems. *Proceedings of the 6th International Workshop on Practice and Theory in Public Key Cryptography – PKC 2003*, Miami, FL, USA, January 6–8, 2003, LNCS Volume 2567, pp. 199–211, Springer Berlin Heidelberg, 2002.

[45] E. Brier, M. Joye: Weierstraß Elliptic Curves and Side-Channel Attacks. *Proceedings of the 5th International Workshop on Practice and Theory in Pubic Key Cryptosystems – PKC 2002*, Paris, France, February 12–14, 2002, LNCS Volume 2274, pp. 335–345, Springer Berlin Heidelberg, 2002.

[46] J.-S. Coron, I. Kizhvatov: An Efficient Method for Random Delay Generation in Embedded Software. *Proceedings of the 11th International Workshop – CHES 2009*, Lausanne, Switzerland, September 6–9, 2009, pp. 156–170, Springer Berlin Heidelberg, 2009.

[47] J. Goodman, A. P. Chandrakasan: An Energy Efficient Reconfigurable Public-Key Cryptography Processor Architecture. *Proceedings of the Second International Workshop – CHES 2000*, Worcester, MA, USA, August 17–18, 2000, LNCS Volume 1965, pp. 175–190, Springer Berlin Heidelberg, 2000.

[48] J. Daemen, V. Rijmen, *The Design of Rijndael: AES – The Advanced Encryption Standard*. Springer, 2002. ISBN 3-540-42580-2.

[49] NIST Computer Security Division (CSD). *SHA-3 Standard: Permutation-Based Hash and Extendable-Output Functions* (PDF). NIST.

[50] D. Hankerson, A. Menezes, and S.A. Vanstone, *Guide to Elliptic Curve Cryptography*, Springer-Verlag, 2004.

[51] E. Candes and M. Wakin, *An introduction to compressive sampling*, IEEE Signal Processing Magazine, vol. 25, no. 2, pp. 21–30, 2008.

[52] A. Fragkiadakis, E. Tragos, and A. Traganitis. Lightweight and secure encryption using channel measurements. *Wireless Communications, Vehicular Technology, Information Theory and Aerospace & Electronic Systems (VITAE), 2014 4th International Conference on.* IEEE, 2014.

[53] S. Gürses, C. Troncoso, and C. Diaz. *Engineering privacy by design.* Computers, Privacy & Data Protection 14 (2011).

[54] A. Pfitzmann, and M. Hansen. A terminology for talking about privacy by data minimization: Anonymity, unlinkability, undetectability, unobservability, pseudonymity, and identity management. (2010): 34.

[55] A. Cavoukian, Evolving FIPPs: Proactive Approaches to Privacy, Not Privacy Paternalism. *Reforming European Data Protection Law.* Springer Netherlands, 2015. 293–309.

[56] D. Chaum, and E. Van Heyst. Group signatures. *Advances in Cryptology—EUROCRYPT'91.* Springer Berlin Heidelberg, 1991.

[57] H. Tschofenig, et al. The IETF Geopriv and presence architecture focusing on location privacy. *Position paper at W3C Workshop on Languages for Privacy Policy Negotiation and Semantics-Driven Enforcement,* Ispra, Italy. 2006.

[58] J. Raymond, *Traffic analysis: Protocols, attacks, design issues, and open problems.* Designing Privacy Enhancing Technologies, 10–29, 2001.

[59] D. L. Chaum, Untraceable electronic mail, return addresses, and digital pseudonyms. *Communications of the ACM*, 24(2), 84–90, 1981.

[60] A. Ruiz-Martínez, A survey on solutions and main free tools for privacy enhancing Web communications. *Journal of Network and Computer Applications*, 35(5), 1473–1492, 2012.

[61] G. Danezis, & R. Clayton, Introducing traffic analysis. *In Digital Privacy: Theory, Technologies, and Practices* (pp. 1–24), 2007.

[62] R. Neisse, I. Nai Fovino, G. Baldini, et al. A Model-based Security Toolkit for the Internet of Things. *International Conference on Availability, Reliability and Security (ARES)*, University of Fribourg, Switzerland, 2014.

[63] R. Neisse, D. Holling, A. Pretschner, Implementing Trust in Cloud Infrastructures. *11th IEEE/ACM International Symposium on Cluster,*

Cloud and Grid Computing (CCGRID), Newport Beach, USA, May 2011.

[64] A. Jøsang, Evidential reasoning with subjective logic, *in: 13th International Conference on Information Fusion*, 2010.

[65] M. Walfish, A. J. Blumberg,: Verifying Computations without Reexecuting Them. Commun. ACM 58(2), 74–84 (2015).

[66] M. Backes, D. Fiore, R. M. Reischuk, Verifiable delegation of computation on outsourced data. In: *ACM CCS*. pp. 863–874. ACM (2013).

[67] D. Catalano, Homomorphic Signatures and Message Authentication Codes. *In: SCN. LNCS*, vol. 8642, pp. 514–519. Springer (2014).

[68] R. Johnson, D. Molnar, D. Song, D. Wagner, Homomorphic Signature Schemes. In: *CT-RSA*. pp. 244–262. LNCS, Springer (2002).

[69] R. Steinfeld, L. Bull, Content Extraction Signatures. In: *ICISC*. Springer (2002).

[70] H. C. Pöhls, K. Samelin, On updatable redactable signatures. In: *ACNS*. pp. 457–475. LNCS, Springer (2014).

[71] C. Brzuska, H. C. Pöhls and K. Samelin. Efficient and Perfectly Unlinkable Sanitizable Signatures without Group Signatures. In *Proc. of the 10th European Workshop: Public Key Infrastructures, Services and Applications (EuroPKI 2013)*, pages 12–30, Springer Berlin Heidelberg, 2013.

[72] H. C. Pöhls and K. Samelin. On Updatable Redactable Signatures. *In Proc. of the 12th International Conference on Applied Cryptography and Network Security (ACNS 2014)*, Springer, 2014.

[73] C. Brzuska, H. C. Pöhls and K. Samelin. Efficient and Perfectly Unlinkable Sanitizable Signatures without Group Signatures. *In Proc. of the 10th European Workshop: Public Key Infrastructures, Services and Applications (EuroPKI 2013)*, pages 12–30, Springer Berlin Heidelberg, 2013.

[74] T. Groß, Certification and efficient proofs of committed topology graphs. In: *CCSW*. ACM (2014).

[75] T. Groß, Signatures and Efficient Proofs on Committed Graphs and NP-Statements. *In: Financial Cryptography and Data Security. LNCS*, Springer (2015).

[76] Y. Chen, V. Paxson, R. H. Katz, What's New About Cloud Computing Security? University of California, Berkeley, Tech. Rep. UCB/EECS-2010-5.

[77] J. Camenisch, A. Lehmann, G. Neven, Electronic Identities Need Private Credentials. IEEE Security & Privacy 10(1): 80–83 (2012).

[78] P. Samarati, k-Anonymity. Encyclopedia of Cryptography and Security (2nd Ed.) 2011: 663–666.

[79] C. Dwork, Differential Privacy. Encyclopedia of Cryptography and Security (2nd Ed.) 2011: 338–340.

[80] D. Slamanig, C. Hanser,. On cloud storage and the cloud of clouds approach. *ICITST 2012*, IEEE (2012).

[81] J. Black, P. Rogaway, Ciphers with Arbitrary Finite Domains. *CT-RSA 2002*, LNCS, Springer (2002).

[82] R. Agrawal, J. Kiernan, R. Srikant, Y. Xu, Order-Preserving Encryption for Numeric Data. *SIGMOD Conference 2004*, ACM (2004).

[83] J. Braun, J. A. Buchmann, C. Mullan, A. Wiesmaier, Long term confidentiality: a survey. Des. Codes Cryptography 71(3): 459–478 (2014).

[84] V. Gagliotti, M. A., Buchmann, J. A., Cabarcas, D., Weinert, C., Wiesmaier, A. Integrity, authenticity, non-repudiation, and proof of existence for long-term archiving: A survey. Computers & Security 50: 16–32 (2015).

7

IoT Analytics: Collect, Process, Analyze, and Present Massive Amounts of Operational Data – Research and Innovation Challenges

**Payam Barnaghi[2], Martin Bauer[8], Abdur Rahim Biswas[3],
Maarten Botterman[9], Bin Cheng[8], Flavio Cirillo[8], Markus Dillinger[7],
Hans Graux[10], Seyed Amir Hoseinitabatabaie[2], Ernö Kovacs[8],
Salvatore Longo[8], Swaroop Nunna[7], Alois Paulin[5],
R. R. Venkatesha Prasad[4], John Soldatos[1],
Christoph Thuemmler[5] and Mojca Volk[6]**

[1]GR, Athens Information Technology, Greece
[2]University of Surrey, UK
[3]CREATE-NET, Italy
[4]TU Delft, Netherlands
[5]Edinburgh Napier University, UK
[6]University of Ljubljana, Slovenia
[7]Huawei, Germany
[8]NEC, Germany
[9]GNKS Consult BV, Netherlands
[10]Timelex, Belgium

7.1 Introduction

Internet-of-Things (IoT) Analytics refers to the process of transforming vast amounts of information from heterogeneous internet-connected objects, data sources and devices (e.g., sensors, appliances, cyber-physical systems, Machine-to-Machine systems) to business and application intelligence. Several tools and techniques for IoT analytics have their roots in conventional web analytics, which process and combine data streams from web-connected computers, cell phones and web databases. However, IoT analytics broaden the scope of web analytics on the basis of the collection, processing and analysis of information produced by internet-connected devices, thus enhancing the scope and functionalities of related applications.

Nowadays, IoT analytics have a growing momentum, which is highly due to the proliferation of IoT devices and the overall momentum of IoT technologies and services. IoT analytics hold the promise to enable a wide range of novel applications that are not currently possible, which could revolutionize applications areas with significant socio-economic impact such as healthcare, energy management, public safety and more. The IoT analytics vision, while fantastic, is associated with several challenges spanning both technical and policy issues. For example, at the technical and scientific forefront, IoT devices tend to produce high-velocity streams, which challenge the capabilities of state-of-the-art BigData systems (such as MapReduce). Furthermore, the heterogeneity and diversity of IoT devices is a serious set-back to the collection, consolidation and unified processing of IoT data streams. Other challenges relate to the selection, refinement and deployment of effective data analytics algorithms that can respond to the stringent QoS (Quality of Service) requirements of IoT applications. Likewise, at the policy forefront, there is a need for addressing security, privacy and data protection challenges in-line with existing regulations, but also in a way that encourages user participation.

The present chapter of the 2015 IERC Book aims at presenting the above-listed challenges of IoT analytics, while at the same time providing insights in possible solutions, notably solutions that are being developed in the scope of the IERC community. The second section of the chapter (following this introductory one), is titled "Deep Internet of Things Data Analytics". It presents challenges and solutions associated with the collection and semantic unification of diverse data streams, which is one of the first prerequisite steps for analyzing IoT data sources. Likewise, the third section of the chapter presents the challenges of IoT/BigData convergence and illustrates techniques for integrating IoT with cloud and BigData infrastructures. The fourth section of the chapter provides insights associated with the practical application of IoT analytics in healthcare and social care, while the fifth section presents an IoT analytics case for public safety. Finally, the sixth section of the chapter deals with the ever important policy issues, through presenting challenges and providing a perspective for solutions that are in-line with existing and emerging EU directives. It also identifies gaps of these directives and proposes relevant remedies. Overall, this chapter provides the reader with a nice overview of the technical and policy issues associated with the wider deployment of advanced IoT analytics, along with some solutions introduced and advanced by the IERC community.

7.2 Deep Internet of Things Data Analytics

7.2.1 Introduction

Computers in their early days were not designed for personal use and individual applications. They were usually large machines and mainframes that specialists worked with. Rapid hardware and software innovations and advancements and the emergence of global networks and the Internet made computers widely available for everyone to use. Mobile devices and wireless technologies made it potentially possible to connect to communication networks and the Internet anytime and anywhere. We now live in an era in which physical objects (i.e. "Things") can be embedded with their own computing devices and with networking capabilities. The Internet of Things (IoT) is an umbrella term that refers to technologies that enable communication and interaction between various devices and real world objects and human users. IoT is mainly enabled by advancements in manufacturing low-cost sensors and actuators, smart phones, embedded devices, and communication and networking technologies. These advancements have resulted in rapid growth and the deployment of networked-enabled devices and sensing and actuation systems that interconnect the physical word with the cyber-world. The number of devices connected to the Internet has already exceeded the number of people on earth and is estimated to grow to 50 billion devices by 2020 [39].

Data collected by different devices are of various types (e.g. temperature, light, humidity, video) and are inherently diverse and dynamic (i.e. the quality and validity of data can vary with different devices over time; data is also mostly location and time dependent). Sensory devices can be ubiquitous and are often constrained in power, memory, processing and communication capabilities. As the scale of interactions between devices and the load of communications rapidly increase, real world data and service traffic become voluminous; the current Internet/Web architecture will not be suited to delivering reliable, efficient and time-sensitive data and services for large volumes of networked devices [1].

In following paragraphs we discuss that IoT data analytics cannot be separated from data collection, device and network conditions and limitations. The ability of the resources to effectively publish, discover and access the data in large-scale distributed environments will have an impact on the efficiency of the data analytics methods. Effective data analytics solutions in the IoT need to consider the dynamicity and constraints of data collection devices and communication networks and should be able to optimise the processes for different purposes and requirements, such as latency, accuracy, data and sampling rates, and energy efficiency. We discuss data analytics in the IoT domain and describe some of the key issues to provide integrated and end-to-end solutions for large-scale and efficient data analytics. The integration of device and network parameters and their characteristics in the IoT data analytics in this work is referred to as *Deep IoT data analytics.*

7.2.2 Designing for Real World Problems

IoT research covers a broad range of technologies and solutions that aim to tackle the challenges in networking and communications, interoperability, services and stream processing and data analytics. IoT is an integration of different systems and technologies. Industry based solutions and services in this domain are often under development or do not interoperate on a global and large scale, due to a lack of standardisation. Data processing and analytics solutions in the IoT are mainly based on conventional data mining and machine learning techniques. There are also several solutions and de-facto standards for annotation and semantic integration of IoT data.

However, IoT data is inherently different from other types of data on the Web and database systems. Uncertainty, incompleteness, sporadic data distributions, scale and energy and resource constraints of the data provider devices are among the key issues that make processing IoT data different and more challenging than the usual data on the Web and database systems.

Data analytics solutions in the IoT, in contrast to many existing BigData analysis works, cannot be separated from data collection, selection, network status and issues such as energy efficiency. Efficient and intelligent data analysis methods for the IoT domain should consider end-to-end and integrated solutions and should be adaptable and flexible enough to work with incomplete and uncertain data and should also be able to adjust themselves to concept drifts (i.e. changes in the data or the objectives of data processing) [2] and requirement changes. The IoT is an online network of resources and data analytics solutions and should be able to process and analyse dynamic data in real-time.

Figure 7.1 (adapted from [3]) shows some of the key dimensions that need to be taken into account when designing data analytics methods for the IoT. As shown in the figure, the connectivity and data publication can potentially be at any time, from any place and can be related to any-thing. The volume of data is a key issue and the networks and communication technologies have an impact on various aspects of data access and use, such as latency, quality and availability. IoT data can be related to people, personal spaces and living

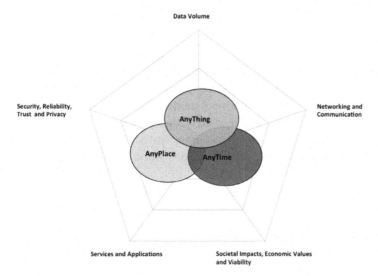

Figure 7.1 Key dimensions in production and deployment of IoT data.

environments; so reliability, security and privacy are among key issues in designing any solutions, including data analytics for the IoT. Having access to new types of data and connectivity and interaction with the real world provides an opportunity to design new services and applications that rely on ambient intelligence. However, data analytics methods for extracting this ambient intelligence have to deal with time and location dependency and dynamicity of data and the solutions should be able to handle uncertainty and quality issues often in a real-time manner. IoT services and applications will have an impact on people's lives and the way that personal and public spaces and services are planned and designed (e.g. smart homes and smart cities). Industrial IoT applications and services require the processing of large volumes of data to make autonomous decisions to control and operate various systems and machines.

Wireless communication is the dominant component in overall energy consumption of the remote IoT devices (i.e. in the current systems the computation usually consumes less energy than the communication [4]). In this regard, careful considerations should be made to minimise the communication load in IoT networks. IoT resources are usually programmed to collect and forward data based on a given data acquisition frequency and are often ignorant of information within the data packets. This could lead to the creation of redundant and unnecessary communication load when data is noisy and unreliable or when it does not contain any significant or new information.

The other important issue is the scalability of the data processing and computation. Recent efforts in distributing processing tasks among different resources (e.g. using software defined solutions [5]), have mitigated the problems associated with conventional centralised processing architectures to some extent. However, with the scale of the IoT resources, problems such as effective use of computational resources on devices and distribution of data analytics processes between IoT devices and Cloud based resources are challenging issues.

7.2.3 Real World Data

To better understand the requirements of data analytics we first need to look at the data sources and the type of devices and networks that produce and handle this data. Access to live real world data and connected worlds of physical objects, people and devices are rapidly changing the way we work and interact with our surroundings and have had a profound impact on different domains, such as healthcare, environmental monitoring, urban systems, industry, and control and management applications and decision support systems.

IoT data is usually collected via sensing devices that are connected to wireless or wired networks (e.g. wireless sensor and actuator networks), smart phones and other embedded and network-enabled devices. The devices can be directly connected to the core network and data analysis components or gateway components can provide data communication between IoT devices and higher-level services and applications, including the data analytics components in the core networks.

Figure 7.2 shows a generic framework for IoT data communication where some nodes can use Internet and Web based protocols and some are connected via gateway components. There are also platforms and solutions that enable crowd sourcing of IoT data collection and publication using smart phone and network-enabled devices and sensing technologies. Quality, trust and reliability, together with the availability and delays in accessing the data are key issues in crowd sourced data collection and publication use-cases.

IoT data is often published as streaming data with multiple streams that can provide similar data (but can have different quality or parameters) or other relevant data that need to be integrated and processed together. Extracting patterns and finding correlations between different parts of the data is an important task in data analytics for IoT data streams. However, there are two

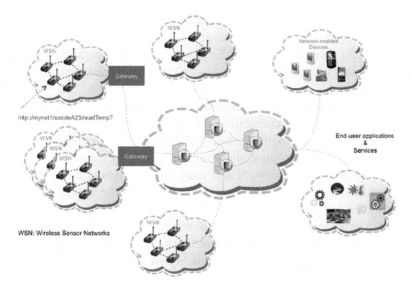

Figure 7.2 A generic framework for IoT data publication and communication.

key issues: causation vs. co-occurrence requires further analysis and often background knowledge is required to interpret and separate the causations; time lag between different pattern occurrences and spatial dependencies should also be considered when analysing the patterns in the streams. For example, an occurrence in a data stream (i.e. an event) can cause a related pattern in a different stream (and in a different location) after a period of time. So the spatio-temporal interdependencies should also be considered in the analysis. The streaming data can sometimes have missing values due to communication and device errors or different sampling rates. Different interpolation techniques (e.g. Gaussian process or multivariate interpolation techniques) or machine learning methods can be employed to compensate for the missing values in the data streams.

7.2.4 Data Interoperability

Data collection and publication is the initial step for accessing and processing IoT data. As discussed, there are several issues regarding the device, network and end-user application and service state and requirements that need to be taken into consideration when collecting and publishing the IoT data.

Data is usually published in various forms and via distributed devices and sources. There are several existing metadata models and description frameworks that are designed and proposed by academia and industry to provide interoperable resource and data descriptions in the IoT domain. For example, the W3C Incubator Group on Semantic Sensor Networks developed a higher-level model for describing sensors and their capabilities called SSN Ontology [6].

These semantic and metadata models and description frameworks are designed to improve the interoperability of the data and resource descriptions. Machine-readable and automatically interpretable data descriptions and data engineering solutions to enhance the structure and representation of data will strongly improve the analytics and integration methods, especially in the IoT world, where multi-modality and heterogeneity are among the key issues. However, the semantic annotation requirements and the complexity of providing structured information with several attributes often hinders the effective use of the semantic models that are proposed for real world data. Some of the parameters, such as quality of data, are also dynamic variables. Most of the current semantic annotation models construct a semantic description model and annotate the data according to that model without providing an end-to-end set of tools and solutions to add and update more attributes and metadata to the annotation after the data is published. However, the dynamicity and changes in the annotated values (e.g. the meaning of quality for a data item can change after the time of measurement; provenance of data can change as more processing methods are applied to the data) is not captured in the models and annotation methods.

Many of the current semantic annotation frameworks for the IoT are static and the provenance and changes to the data and metadata updates are not directly supported. Providing "dynamic semantics" in the IoT domain and developing tools, APIs and methods that can publish, update and extend the semantics as the data is processed and integrated with other sources, or as more information is collected and analysed from the environment will help to resolve this issue. This will not only address the interoperability issues but will also create more enhanced and flexible annotations that reflect the actual and up-to-date attributes of the data. The dynamic semantic methods should also use linked-data descriptions to link between different resources and also use common vocabularies to describe the concepts and content of the annotations. Using common vocabularies and topical ontologies for describing events and occurrences and other common attributes of the data, such as units of measurement, will significantly improve the interoperability and effectiveness of data analytics operations.

7.2.5 Deep Data Analytics Methods

Edge-level pre-processing, filtering the noise and removing the corrupted data and data aggregation mechanisms on the IoT device could help to minimise the use of communication resources at source level. Pre-processing and data aggregation at device level is a remedy for the congestion problem that often occurs in centralised and hierarchical architectures and will lead to a more scalable design.

In the IoT, data analysis algorithms should be able to automatically make adjustments and adapt the overall solutions to different information extraction and optimisation goals. For example, in an emergency response scenario the algorithms need to be optimised for reducing latency; in an elderly care scenario increasing the quality of the extracted information would be the main priority; in an environmental monitoring framework using a large number of wireless sensors and increasing the life-time of the network can be one of the main goals of the overall application and consequently the data analytics method should also be adjusted and optimised to meet these goals and requirements. Obviously IoT devices in large deployments will not run for just one application and will not respond to a single demand, so cross-application optimisation is also an essential task in developing large-scale and multi-purpose IoT frameworks. To perform such optimisation and integrated data processing efficient data discovery and selection algorithms for choosing the best set of resources at the given time, and adaptable and customisable data analytics methods that can push the processing to the edges of the IoT networks, are required.

Most of the conventional machine learning and data analytics methods are also designed based on the assumption of having normalised distribution and reliable datasets. For example, processing techniques, such as the Symbolic Aggregate approXimation (SAX) [7] algorithm for constricting patterns from streaming data, assume that the data distribution are normalised. SAX divides the normalised probability distributions to equi-probable segments and assigns a symbolic representation for each segment. These symbolic representations are then used to create representative patterns of the streaming data. While SAX or other similar methods have been used effectively in the data analysis and stream processing fields, using them in the IoT has some limitations. SAX patterns can still be constructed from the IoT data streams but the distribution of the IoT data in short-time windows, which is often the key focus of the (near) real-time data analytics methods, is not normalised and can be a sporadic and multivariate distribution. This will require pre-processing and analysis of the data at source level and determining the distribution and other attributes of the data before constructing the symbolic representation and constructing the patterns.

Data analytics for dynamic environments such as IoT requires resource-aware analysis techniques that focus on both the data and also the resources that provide the data. Optimisation for different objectives such as latency, accuracy, energy efficiency, and network traffic should be supported and the algorithms should be able to adjust and adapt to these objectives dynamically. The key target of data analytics in the IoT is to create situation-awareness and ambient intelligence and to extract actionable information that can be used in decision support systems and higher-level applications and services. The results of the data analytics can also be used to visualise and demonstrate different patterns, occurrences and events in the physical environments. Most of these are online applications and services that require (near) real-time learning and feedback mechanisms.

Figure 7.3 shows a multi-level view of data analytics in an IoT framework. The device and resources are the edge-level and their parameters at any given time will have an impact on real-time data collection and other parameters, such as quality and granularity of the data. The analytics methods need to take into account these parameters and to also try to control and adjust these using

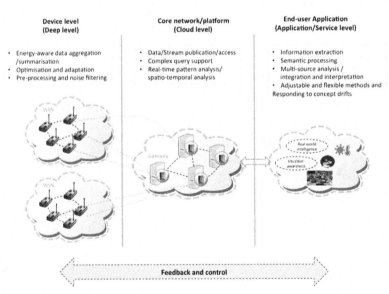

Figure 7.3   A view of data analytics levels in an IoT framework.

software defined and adjustable solutions to provide more resource-aware solutions. The middle layer is the core network and Cloud based services that can provide back-end support for discovery, integration, publication and storage, and large-scale distributed analytics methods. The functions at this level will be adapted according to requirements, concept drifts at the end-user and application/service layer and also condition and priorities at the device level.

In the machine learning domain and the Big Data world there are deep learning methods that attempt to learn representations and model abstractions of data [8]. The deep learning methods often also improve the performance of the learning methods by analysing and processing large volumes of data. In the IoT domain, the use of deep learning and other conventional and novel data analytics and stream processing methods can be very beneficial. However, the deep analytics, as described in the paper, mainly describes the adaptability and adjustability of the methods towards various optimisation objectives and concept drifts and is an attempt to develop analytics and machine learning techniques that can take device and network parameters into account and can work efficiently with multivariate and sporadic data provided by multiple sources and by various qualities.

The data analytics solutions in this IoT domain also rely on semantic annotations and descriptions of the resources. The more expressive the attributes of the data and their provider resource are, the better the interpretation and analysis that can be provided. However, expressive semantic annotations and metadata will require a higher communication and computation load (and consequently will consume more energy in constrained environments). The update, query and processing of complex semantics can also their hinder efficient utilisation. So a trade-off between semantic descriptions, efficient publication, query and discovery methods and adaptable and flexible learning and analytics solutions are required in the IoT framework.

7.2.6 Conclusions

Increased interest in using the IoT in different domains, such as smart cities, healthcare and industry, plays a key role in the production of massive amounts of real world data. This data is mainly collected in order to extract actionable information, create ambient intelligence and provide situation awareness for different higher-level applications and services [9]. IoT is a dynamic environment with various devices that are often resource constrained and deployed on a large scale. Consequently, IoT data is also dynamic and heterogeneous.

In contrast to many Web and database systems, data analytics methods depend on the context of the data production source and network and device parameters.

Efficient IoT data analytics methods require end-to-end techniques and solutions for collection and publication, discovery and selection, and adaptable and adjustable data analysis mechanisms and techniques. Drifts and changes, both in the end-user targets and operational environments and optimisation goals and network and device parameters, should be monitored and captured and should be fed back to control mechanisms that can adapt and control the data analytics methods. Key challenges for the future generation of IoT data analytics, in addition to overcoming the scale, computation and multi-modality issues, is to provide software controlled and adaptable solutions that can monitor the changes deep in the networks and physical environments and optimise their functions and goals based on end-user requirements, network and platform context, and changes in the surrounding environment.

7.3 Cloud-Based IoT Big Data Platform

7.3.1 Introduction

The third generation Internet of Things (IoT) comprises millions of applications, billions of users and trillions of devices. Over the last years, IoT has moved from being a futuristic vision to market reality. It is not any more a question that whether IoT will exist surpassing the hype, but it is already there and *IoT industry* race has already begun. Trillions of connected devices are the enablers; however the value of IoT is in the data and advanced processing of the collected data. IoT data is more dynamic & heterogeneous, imperfect & unstructured and, unprocessed & real-time than typical business data. It demands more sophisticated IoT-specific analytics to make a meaningful inference. The exploitation of the real-time big data obtained from sensors/actuators in IoT context by processing in sophisticated cloud is very much a necessity. This data processing leads to advanced, proactive and intelligent applications and services. The colligation of IoT and Big data can offer: i) deep understanding of the context and situation, ii) real-time actionable insight – detect and reacted to in real-time, iii) performance optimization, and iv) proactive and predictive advanced knowledge. Cloud technologies offer decentralized and scalable information processing and analytics, and data management capabilities.

Following paragraphs address the cloud based IoT and Big data platform concept and their emerging requirements on the convergence of sensors and devices, big data analytics, cloud data management, edge-heavy computing, machine learning and virtualization. Initial results from iKaaS (Intelligent Knowledge as a Service) an EU-Japan project on IoT/Cloud/Big data are also discussed.

7.3.2 Big Data in the Context of IoT

Big data is defined by 4Vs (Figure 7.4), these are Volume-, Velocity, Variety, Veracity. Volume means large data size in 100s of terabytes. Velocity means the real-time and/or stream of data. Variety means the heterogeneous data (e.g., structure and unstructured, diverse data models and query languages, and diverse data sources). Veracity means data uncertainty due to data inconsistency, incompleteness, ambiguities, latency, model approximations, etc.

IoT faces all 4Vs of the Big Data challenges. However the velocity is the main IoT Big data challenge because of real-time and stream of data coming from diverse IoT devices and sensors. Real-Time Big Data terminology is often replaced by the term IoT Big Data. The data coming from the IoT devices have to be processed in real-time to arrive at reliable and intelligent decision. For example, healthcare wearables (like ECG (Electrocardiogram) devices) produce up to 1000 events per second which is a challenge for real-time processing considering miniaturized devices and number of such devices. Next is the volume, for example, large scale IoT deployments gather and process millions pieces of data from millions of sensors per day. Likewise, a wearable sensor produces about 55 million data points per day.

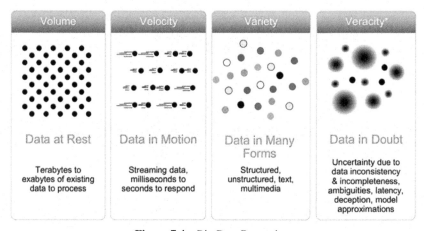

Figure 7.4 Big Data Properties.

7.3.3 Applications of IoT Big Data Analytics

The applications of IoT Big Data analytics can be classified into five main categories which are depicted in Figure 7.5 and include:

- Predictive analytics,
- Prescriptive analytics,
- Descriptive analytics,
- Monitoring and
- Control and optimization.

All these require a deep understanding of the domains, situation and the requirements of services by users.

Gaining insights and knowledge in real time and actionable insights can lead to performance optimization. All the above five applications are inter-related and requires multiple tools like machine learning, reasoning, optimization, etc.

Predictive analytics is used in many applications where users require services that can foresee the situation and act on it. Prescriptive analytics can provide many possible actionable decisions and also can provide the trade-off between them.

Descriptive analytics offers the insights into the situation and helps in deep understanding. Monitoring, control and optimization are legacy applications, but with big data analytics they can be improved immensely.

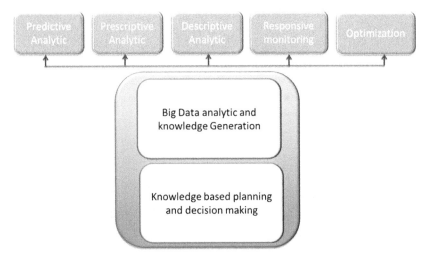

Figure 7.5   IoT-Big Data Applications.

Thus, analytics can indeed offer multiple services such as observing behaviour of things, gaining important insights and processing in real time for immediate actions. For example, in healthcare services IoT analytics can be used for understanding the cause of diseases, as well as for identifying emergency situations.

This vision boils down to solving multiple challenges: to store all the events (velocity & volume); to run queries over the stored events; (velocity & volume) to perform analytics (data mining and machine learning) over the data to gain insights. Examples include real-time fall detection and potential reactions for aging population. Real-time detection and action represent multiple challenges.

7.3.4 Requirements of IoT Big Data Analytic Platform

An IoT BigData Analytics Platform is a real-time online platform that dynamically manages IoT data/objects but it also provides connectivity to the diverse heterogeneous objects, considering the interoperability issues. Next is deriving useful information and knowledge from this connection and large volume of IoT data. The platform needs to offer ubiquitous accessibility and connectivity in facilitation of maximum accessibility as well as connectivity of the diverse heterogeneous objects/services and various volumes of users including mobility. Dynamic management/orchestration of users, billions of devices as well as massive amount of data produced by those connected devices, maximum resource utilization, and sharing of IoT resources (objects, applications, platforms) are all necessary. Personalized, secure, and privacy by design services based on preferences of users and requirements including real-world context are the important requirements. Some of the requirements are briefly discussed in following paragraphs.

7.3.4.1 Intelligent and dynamic

The platform should include intelligent and autonomic features in order to dynamically mange the platform functions, components and applications. The platform should also be capable of making a proactive decision, dynamic deployment, and intelligent decision to understand the context of the environment, users and application requirements, etc. Considering performance targets/constraints, offloading from clients/hosts to cloud is necessary but the performance should be guaranteed. Dynamic resource sharing and service migration is a must for large scale IoT applications. Dynamic metering may be also necessary when IoT devices are shared.

7.3.4.2 Distributed

The platform should include distributed information processing and computing capabilities, distributed storage, distributed intelligence, and distributed data management capabilities. This need to be distributed across smart devices, gateway/server and multiple cloud environments. More distributed processing and storage of the massive data as well as cloud functionalities is a must. Decentralized (and infrastructure-less) clouds will be the order of the day through processing capabilities and positioning data closer to users.

7.3.4.3 Scalable and elastic

The platform has to be scalable to address the connectivity from small to large number of the devices, manage the different scale of the data and services, as well as users. Cloud and edge data management, storage and processing, need to be scalable and at the same time elastic.

7.3.4.4 Real-time

Real-time data processing and service provisioning of "Big data", is necessary. Un-structured and semi-structured data coming from distributed sources should be processed to provide real-time/near real-time services.

7.3.4.5 Heterogeneous (unified)

Interoperability between cloud/IoT services and infrastructure, and federation between cloud, Big data and IoT devices has to be in place to realize full potential. Standard APIs to deal with heterogeneity need to evolve. Open software components, standard data structure and modeling and abstraction of heterogeneous IoT devices and the data is necessary. Data always raise heterogeneity problems: many data formats, many metadata schema descriptions, mix of various levels of complexity, etc., are the cases in point. The target is to deliver a data model and the specification of required mechanisms for exploiting both structured and unstructured data, for moving from raw data to linked data, enabling the adoption of a common understanding, the recognition of similar data, and unambiguous description of relevant information for multimodal and cross-domain smart space applications.

7.3.4.6 Security and privacy

Security and privacy by design is also needed including different privacy and security features like data integrity, localization, confidentiality, SLA (Service Level Agreements), security and privacy-preserving data management modules. Holistic approaches are required to address privacy & security issues across value chains including privacy by design aspects, software algorithms and new data management models.

7.3.5 Cloud-Based IoT Analytic Platform

The cloud-based platform is dynamic in nature and offers flexible resources sharing and service provisioning. It also offers a scalable and elastic service/resources management platform. The platform also offers reliable and easy access to the services using large amount of computing and storage resources. The cloud-based platform is also homogeneous (unified) which reduces the technological heterogeneity. On the other hand, IoT depends on massive resources available when needed and scaled back when not needed. This can only be achieved using cloud paradigm. For IoT, cloud computing functionalities enable the realization of the IoT vision. For Cloud, IoT provides huge opportunities for cloud services. There are two basic approaches for the convergence of IoT-Big data and Cloud. These are (i) Cloud-centric IoT (bringing IoT functionalities into Cloud) and (ii) IoT-Centric Cloud (bringing Cloud functionalities into IoT).

In the following, we provide an overview of the iKaaS platform that has been developed as an example and which is illustrated in Figure 7.6. It combines ubiquitous and heterogeneous sensing, along with big data and cloud computing technologies. iKaaS enables IoT processing consisting of continuous iterations on data ingestion, data storage, analytics, knowledge

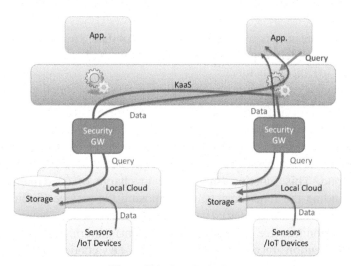

Figure 7.6 iKaaS Platform.

generation and knowledge sharing phases, while at the same time providing a foundation for service provisioning. The iKaaS platform comprises of the cloud ecosystem that consists of *Local Clouds* and a *Global Cloud*. More specifically:

- A local cloud is created on-demand; it comprises of sufficient/appropriate computing/storage/networking capabilities, and provides requested services to users in a certain geographical area and time period as well as offers additional processing and storage capability to services.
- The global cloud is seen in the "traditional" sense, as a construct with on-demand/elastic processing power and storage capability. It is a "backbone infrastructure", which increases the business opportunities for service providers, and the ubiquity/reliability/scalability of service provisioning. It offers more opportunities for offering services, more options on which service features are based in case of context changes, more resources for deriving meaningful decisions, and elastic provision of resources on demand.

Local clouds can involve an arbitrarily large number of nodes (sensors, actuators, smart-phones, etc.). The aggregation of resources comprises sufficient processing power and storage space. The goal is to serve users in a certain area. In this respect, a local cloud is the virtualised processing, storage and networking environment, which comprise IoT devices in the vicinity of the users; users will exploit the various services composed of the devices in local clouds and their capabilities e.g., a sensor and its gateway equipped with the iKaaS platform.

The global cloud can enable, as a special (yet important) case, the existence of IoT service providers capable of providing larger scale services without owning actual IoT infrastructure.

The Cloud ecosystem comprises the following essential functionalities:

- Consolidated service-logic/resource descriptions/registries as part of the Global Cloud enabling the reuse of services. Practically, a set of registries will be developed enabling the pooling of service logic and resources.
- Autonomic service management, first the global cloud and then, in the local cloud. This functionality will be in charge of (i) dynamically understanding the requirements, decomposing the service (finding the components that are needed); (ii) finding the best service configuration and migration (service component deployment) pattern; (iii) during the service execution, reconfiguring the service, i.e., conducting dynamic additions, cessations, substitutions of components.

- Distributed data storage and processing is anticipated for global and local clouds. This means capabilities for efficiently communicating, processing and storing massive amounts of, quickly-emerging, versatile data (i.e., "big data"), produced by a huge number of diverse IoT devices.
- Derivation of information and knowledge (e.g., on device behaviour, service provision, user aspects, etc.), while ensuring security and privacy as a top concern.
- Knowledge as a service (KaaS) will be primarily part of the Global Cloud. This area covers: (i) device behaviour aspects; (ii) the way services have been provided (e.g., through which IoTresources) and the respective quality levels; and (iii) user preferences.

Thus the iKaaS functionality will determine the optimal way to offer a service. For instance service components may need to be migrated as close as possible to the required (IoT) data sources. IoT services may need generic service support functionality that is offered within the cloud, and, at the same time, they do rely on local information (e.g., streams of data collected by sensors in a given geographic area), therefore, the migration of components close to the data sources will help in reduction of data traffic.

7.4 IoT Analytics in Health and Social Care

7.4.1 Introduction

Following a protracted start back in the early 2000, IoT is nowadays an undeniable force, which will dictate our (virtual) reality in years to come. According to figures by Cisco the global amount of mobile data will grow dramatically to an annual run rate up from 30 exabytes in 2014 to 249 exabytes by 2019 [10]. While a significant share of this data, almost 79% will account for IP video streaming by 2018 it is reasonable to assume that by then 5–10% of the overall traffic will be generated by smart devices, sensors, attenuators, embedded – and cyber-physical systems. Although these developments are currently driven in the first instance by industrial domains such as automotive, retail and logistics there is evidence for massive utilization of IoT strategies in the health and social care domains in the near future.

It has become increasingly clear that the way health and social care will be delivered in the future is undergoing substantive changes. These changes are driven by the demographic and socio-economic developments in our societies and also the technological and bio-medical progress. As a general trend the availability of smart IoT capable devices has dramatically improved. Wearables are everywhere, from smart glucometers for blood-glucose measurement, insulin-pumps and highly complex brain-pacemakers for the treatment of Parkinson's disease to the "iWatch" or similar products. Portability has increased since storage capacity of smart phones has reached hundreds of Gigabytes and the battery capacity is significantly improved.

Governments in Europe are now publically debating the utilization of IoT technology to control health and social care costs by enabling and empowering patients and their informal carers [11]. However, while the focus of IoT research has so far been placed on creating reference architectures and conduct design- and feasibility studies in order to interlink devices and capture and collect information the issues around analytics and the creation of value are now taking center stage [12].

Even though the focus of this section is clearly on analytics and enabling architectural designs it is certainly important to underline that in sensitive areas (such as healthcare), the discussed technologies and their possibilities have to always be set into perspective with the relevant ethical and legal considerations [13]. For further information on this topic the interested reader may wish to consult the ethics, science and technology section at the European Political Strategy Centre [14].

7.4.2 Architectural Approach to Data Analytics

Essential to the health – and social care domains is the understanding that architecture should be scalable and able to cater for the analysis of big and small data whereby the topology is becoming more and more relevant [15, 16]. Conventional cloud computing has long been regarded as the holy grail of big data analysis and is certainly a powerful method. The ability to share computing resources and balance the load according to the need of the task at hand makes cloud infrastructures clearly a formidable approach. Typical examples include the analysis of pre-existing large databases such as censuses or genetic information (genomics) databases.

However, standard cloud approaches seem to struggle with some require-ments of the health and care domains, especially with regards to time critical processes. Although cloud approaches are powerful strategies, the bottle-neck seems to be the network and the relatively high latencies associated with it. Furthermore, there are continuous privacy and security concerns associated with public clouds for the use in health and care.

The biggest challenge seems to be the fact that the predicted growth of network traffic, especially mobile traffic, will outperform the network capacity by 2019 [10]. At the same time there is clear evidence for a significant increase in sensors, attenuators, embedded and cyber-physical systems, which on the one hand clearly drives the utilization of IoT technology in e-Health but also drives the increase in data, which further widens the gap between traffic demand and network capacity. This dilemma has caused a paradigm shift towards a distributed analytics approach, which might be the way forward in the health and care domains, which are set to generate very large amount of data with the potential to jam up existing infrastructures.

While hybrid-cloud models were early manifestations of the attempt to solve the privacy problem in sensitive areas such as health and care, most recently this has been developed further into more sophisticated strategies involving mobile edge cloud computing and lately the so called "fogging" [16–18]. Fog computing in a way merges the benefits of cloud computing and grid computing as it on the one hand integrates peripheral smart devices in one distributed approach while on the other hand allows for local problems to be solved locally. This has implications with regards to latency, privacy, precision, autonomy and liability [19, 20].

While politicians and administrators still push for electronic patient files or electronic health records there is an urgent demand to clarify the terminology. As it is unlikely that a homogeneous data base system ideal for the assessment through classic cloud strategies can be achieved in most European countries in the foreseeable future hyper-distributed models where patients will be using their own smart devices to collect and manage their own data will become the norm. Fogging, supported through mobile edge clouds might be far superior to conventional clouds in such a scenario. New hyper-distributed architectures could also protect clinical infrastructures from being over-loaded with irrelevant information while it allows for patients and informal carers to be in full control of their information. It will also protect health care providers from the risk of loss or theft of information and reduce their exposure to litigation.

7.4.3 IoT Data Analytics

Big data analytics in healthcare is considered a transformational science, which has gained much attention in recent years. Doubtlessly, this can be attributed to unsustainable costs in healthcare, which calls for IT-assisted solutions. Growing adoption of patient-centred mobile digital health applications, availability of advanced cloud and connectivity options, and the rise of the wearables allowing for continuous observation of health-related events have already massively increased the amount of health-related information. Known as mobile or pervasive healthcare, remote collection of personal health and environmental data through sensor networks and mobile devices is well underway. This creates opportunity to track healthy behaviours, understand diseases at a granular level and provide true patient-centred care, and might fundamentally alter healthcare services as the industry moves to value-based models.

In order to exploit and fully leverage this long-term potential in the context of healthcare, advanced big data analytics are needed to enable extraction of valuable and actionable insights and establish sustainable value chains [21, 22]. The use of analytics solutions in healthcare is being increasingly recognized for its value in delivering quality care and gaining competitive edge. Most importantly, tools are needed to cope with the 4Vs [23] of Big Data in healthcare (i.e. Volume, Velocity, Variety and Veracity) and to translate "noisy masses of data" into unambiguous, quality and meaningful insights that can be safely applied with confidence to practice on both patients' and experts' sides. In turn, the healthcare analytics market is growing at a rapid pace, and there are several good practice examples of use resulting for example in lower hospital readmissions and shorter hospital stays, and successful monitoring and prevention of chronic diseases. According to MarketsAndMarkets [24], the healthcare analytics market is experiencing substantial growth at a Compound Annual Growth Rate (CAGR) of 25.2% and is expected to reach $21,346.4 Million in 2020. However, obviously the potential comes with a price – expectations are high and requirements are strict around security, privacy and protection of sensitive information and establishment of trust is necessary throughout the value chains. If not addressed appropriately, these might present the biggest barriers for adoption.

In terms of research and science, Big Data is a well-developed field when it comes to principles, algorithms, methods and tools for data collection, cleaning, description and interpretation. The presently established descriptive analytics in healthcare are giving way to predictive and prescriptive techniques to process volumes of heterogeneous messy data harvested from various sources and integrated across distributed infrastructures. Semantic science, machine learning and classification mechanisms provide for powerful interpretation and translation techniques, for example to integrate and quantify sources with insights into patients' personal point of view, such as Twitter or self-reporting mobile apps, and translate subjective observations into objective medical terms. Other recognized techniques are also statistical analytics, fact clustering, and natural language processing. However, regardless of such advanced techniques, data analysis is frequently the application's bottleneck, both due to insufficient scalability of the underlying algorithms and due to the increasing volume and complexity of the source data, which is continuously challenging current approaches. This, in addition to data processing and interpretation science, opens also a whole new avenue of research related to capabilities, capacities and coping strategies when transmitting masses of healthcare data. In this respect, the rise of cloud computing has introduced dramatic shifts in how data is processed flexibly, efficiently and in a scalable way over distributed architectures and shared resources. The cloud computing market for healthcare itself is expected to reach $5.4 billion by 2017, according to MarketsAndMarkets [25], whereas the concepts of Data as a Service (DaaS), Software as a Service (SaaS), Platform as a Service (PaaS) and Infrastructure as a Service (IaaS) are examples of already highly adopted cloud services for bioinformatics data processing. This drives further research in various areas, which is now looking for example into declarative approaches for expressing programs to achieve transparency and optimizations of large and heterogeneous cloud clusters on a global scale. Another research direction focuses on new communication technologies, such as Software Defined Networking (SDN) and Software Design Data Centres (SDDC) intended to support the massive increases in Internet bandwidth and complexities introduced by IoT, which extend beyond bandwidth requirements and device count, such as lower latency, greater determinism and processing closer to the edge of the network [26]. The latter, known as fog or edge computing, is a step further towards coping with bandwidth and latency constraints as well as to support scalable distributed big data analysis using context-aware localized computing.

In essence, fog computing capitalizes on the proliferation of smart devices with increasingly powerful processing capacities and moves some of the transactions and resources from the centre of the cloud to its edge and inventively reuses processing capacities of existing devices rather than establishing channels for cloud storage and utilization [27]. This aggregates selected data at a certain access point and localizes selected processes, hereby reducing the need for large bandwidth capacities on the cloud channels, processing delays and enormous data management capacities at central locations, and finally leading to improved efficiency and reduced costs. This approach is highly promising for IoT in general and including healthcare, and seems to be particularly well-suited for applications for which cloud-based approaches might be either less suitable or less feasible, for example applications that are latency-sensitive, highly distributed in geographic terms or fast-operating in near-real time, especially in Health 4.0 applications [16]. In addition, fog computing keeps the data at its source without sending it into global networks, which presents another crucial benefits for healthcare, namely facilitating the software-to-data paradigm, which is recognized as the approach to be taken in healthcare to better cope with the security, privacy and data protection requirements as well as to reinsure the users about where their privacy-sensitive data is located [17]. Current approaches suggest the use of machine learning models to support the training process taking place on a fraction of data in the cloud, followed by localized and highly optimized data processing on a resource-constrained smart device, for example smartphone or embedded device, using techniques such as decision trees, fuzzy logic or deep belief network [28]. These new avenues of research, hand in hand with the rising 5th generation of telecommunications networks (5G), represent promising advancements towards transformational patient-centred and quality-driven healthcare for the future.

7.4.4 IoT Data Governance and Privacy Implications

Along with increasing computerization tendencies towards "informated healthcare", focus shifts on novel issues such as the governance of data ownership, data access control, accountability, security, and privacy. In a nutshell, challenges arise on questions such as who has access to the collected and stored data, how is it anonymised and/or de-personalized, and how non-repudiation of data exchange, possession and creation can be assured, which is a crucial prerequisite for the integrity and trust in the data at stake.

Healthcare data is very specific data – unlike data which is collected for traditional purposes of e.g. commerce, transport, logistics, or control over manufacturing processes, healthcare data is a special kind of personal data, which is subject to detailed legal regulations, policies and jural decisions. Data used in the healthcare domain is often so-called personal data, which is a legal term denoting (1) any information, which is (2) relating to (3) an identified or identifiable (4) natural person. This legal concept is deliberately kept rather broad and lacks clear and direct applicability for information systems developers. There is a however a need to separately clarify whether or not a piece of data has to be considered personal data under the respective regulations. A good and substantiated overview on what constitutes personal data with regard to the EC Directive 95/46/EC (Data Protection Directive) and Directive 2002/58/EC (E-Privacy Directive) is provided by Opinion 4/2007 of the Data Protection Working Party [29], nevertheless, legal assistance might be advisable in order to determine how to treat data properly.

Aside from the particularities emerging from legal data, system designers and developers must take into consideration that access to the thus collected and stored data might be requested by multiple heterogeneous stakeholders. The data subject, i.e. the person, who the data is about, is entitled to know which data is collected and to receive access to the collected data, to demand its rectification, and in certain cases, its destruction. Aside from the data subject, access to the data in the healthcare domain can be requested by third parties for reasons of research, disease prevention/control, and for other purposes of governance bodies. Access to the collected personal data thus can be requested by a set of stakeholders with justified interest, which cannot be fully foreseen at design time of the information system. A further level of complexity is introduced, as the data subject is eligible to know with whom the data has been shared and who is in possession of its data, in order to demand deletion/rectification of the data. The resulting constraints imply new demands to information system designers and developers, who need to take into account complex requirements, which might unforeseeably change in the future due to interventions by law [30].

In order to accommodate for these constraints, new principles of data governance have been introduced in the past years, most importantly the concept of fine grained access control (FGAC), and fair non-repudiable message exchange (FNR). FGAC refers to the ability of databases to govern access to core data based on access policies, which take into account the contents of the data query, the context of the request, and the identity of the requester. Unlike with traditional approaches, which categorize access permissions based on the pre-assigned role of the requester, FNR does not rely on roles, but rather on the complex context. Technologies for FNR have been described e.g., in [31] and [32] which focus on FNR technologies that utilize SQL query rewriting for governing access to the data. Standardization efforts have been conducted by OASIS, which provides the eXtensible Access Control Markup Language (XACML), while IBM introduced the Enterprise Privacy Authorization Language (EPAL). XACML plays an important role also in the European Future Internet landscape, where a dedicated FIWARE Generic Enabler module aims to provide XACML to the IoT domain.

Fair non-repudiation (FNR) reveals its utility when personal data between two entities must be exchanged in such way, that the exchange cannot be refuted by any of the participating parties. This way, a non-repudiable trace chain is coined, which then can be accessed by parties with a vested interest. A state-of-the-art summary on FNR has been provided in [33], where an internet-based approach is described for the direct message exchange between two technical systems.

Combining FGAC and FNR enables the creation of sustainable technology for the governance of data in domains of healthcare, public governance, or public education, i.e. in domains, where governance of access to personal data is subject to public domain policies and influenced by the legal domain.

7.5 IoT Analytics for Public Safety

7.5.1 Introduction

Today Internet of Things (IoT) technologies are transforming our living space into intelligent Smart Environments (Smart City, Smart Home, Smart Building etc.). Smart Environments are equipped with a variety of sensors for capturing information and analysing data in a 'Smart Way', extracting actionable insights and adapting their behaviour to the needs of the users. With the number of connected IoT devices growing into the billions – e.g., Cisco forecasts 50 billion devices connected by 2020 [39] – IoT analytics start to become more and more popular because only raw sensor data are not sufficient to deliver the right QoS to the users. Only if we can give a meaning to IoT data and extract the relevant information on the right abstraction level, the Internet of Things vision can become reality.

7.5.1.1 IoT analytics

Under the constraints of IoT system and the requirements from IoT applications, analytics are playing an important role in the information lifecycle of IoT. Ultimately, IoT analytics enables to find the relevant piece of information in the flood of IoT data, identifying the anomalies that require attention, extracting the unknown patterns and helping to predict what is going to happen next.

IoT analytics need to deal with the IoT system characteristics where the data are highly heterogeneous, dimensional, and unstructured, coming from various data sources even in different business domains. This creates new challenges in the analytics area where problems like data distribution, data reliability, real-time data processing and many others need to be addressed. In addition to that, IoT applications are also expecting a certain quality in the provided data, ranging from insightful statistic results and meaningful patterns for making planning and optimization to real-time predictions and suggestions for making timely or even automated decisions.

Smart cities are a good example of large-scale IoT systems where IoT analytics is highly demanded with great potential to make benefits. For example, there are about 12,000 sensors deployed in the city of Santander [34] that provides information about environmental conditions, parking availability, traffic density, weather and irrigation information. Therefore, in today's Smart Cities, there is already a large quantity of information, but by applying more advanced IoT analytics, more relevant information can be extracted.

7.5.1.2 IoT analytics for public safety
In following paragraphs, we explain the challenges of IoT analytics using Public Safety as the application domain, which is one of the most important aspects of a Smart City. Based on the results delivered by advanced IoT data analytics, we cannot only make city planning and operation smarter, but we can also improve and ensure the safety of citizens [35]. As the sensors deployed in Smart Cities monitor the city pulse and report various situations all around our cities in an 24×7 basis, potential safety problems can be identified early and be localized better, therefore effective actions can be taken in time to improve the safety and well-being of the citizens. Many studies show that, even with cheap but widely deployed sensors, important safety issues can be identified early and swiftly addressed, e.g. the formation of a crowd of people, the breakout of a fire [40], a burst pipe [41] or a blocked street.

One of the challenges for IoT analytics to enable Public Safety is to be able to sense and react to critical situations and mine raw sensor data in real-time. This is mainly because the raw sensor data are very noisy, heterogeneous, and high dimensional, which introduce many complexity and computation difficulties to extract high quality results in real-time. To address this challenge, the following technical problems have been taken into account in our IoT analytics solutions for Public Safety.

- *Establishing dynamic communication channels*: in a typical IoT system like Smart Cities, IoT data flow from sensors to various analytics applications and then actionable results are derived. Automated actions are requested from deployed actuators. The first problem to be solved for IoT analytics is to establish communication channels among sensors, analytics applications, and actuators, in a dynamic, flexible, and scalable way, so that information flow between different components can be easily ensured.

- *Dealing with big data in real time*: Real-time is a very important aspect regarding Public Safety, because critical situations need to be detected immediately or to be predicted early enough. In this case, authorities will have enough time to take actions to avoid potential safety problems. For example, an algorithm of Crowd Detection must be fast enough to identify an emerging crowd situation and then trigger an alarm to inform authorities. To reduce the latency from generating raw data to taking actions, the following issues must be considered by IoT analytics: 1) how to control the frequency of data generation; 2) where to do data pre-processing; 3) how to design algorithms for parallelized real-time data stream processing; 4) how to orchestrate resources in the cloud and at the edge to do scalable data processing.

- *Achieving actionable insights with good accuracy*: results derived from IoT data must be actionable, meaning that the results are understandable, accurate and timely enough to allow authorities to make effective actions. If the results of Crowd Detection come one hour after the crowd event happens or most of the detected crowds are false positive, this type of analytics is not usable for enhancing the Public Safety. Therefore, efficient and advanced machine learning or prediction algorithms must be used by IoT analytics to provide real-time feedback to the authorities.
- *Preserving user privacy*: as data are collected from different sources, one obvious issue is the user privacy. Privacy protection and governance must be seriously taken into account from the start. This will affect the choice of our solutions.

Of course, data and system security is another technical issue for IoT analytics, but it is not regarded as a key focuses of this paper. In the remainder of this paper, two specific solutions are introduced to explain how we improved Public Safety via IoT analytics for outdoor and indoor use cases.

7.5.2 Crowd Detection Solution for a Safer City

Efficient emergency systems require a number of different technologies to monitor and detect dangerous events in real-time. Several problems arise in the design, implementation and development of such systems. One of the main problems that affect such systems is human behaviour in critical situations. Being able to detect dangerous situations and act in real-time is a need for enhancing people's safety but is not an easy task due to the variety of the human behaviours. In this case, IoT analytics can help to fuse and mine sensor data from various installations to produce actionable insights. One example of IoT analytics' solution is the privacy preserving Crowd Detection component from NEC.

Concretely, the Crowd Detection core functionality is on understanding the dynamics of crowd. This requires in-depth understanding of how humans move in an indoor space over time. Most state of the art approaches are using video based crowd analysis. They face deployability issues on account of privacy regulations and the public's perception of surveillance. Our approach is based on privacy preserving sensors, guaranteeing that no collected information can be used to identify an individual.

In addition to the citizens' safety, our crowd analysis can provide relevant information for the design of public spaces, e.g., making shopping malls more comfortable for customers, enhancing their safety, or coordinating evacuation plans based on the real-time crowd behaviour. The Crowd Detection solution can also be used for automated detection of anomalies and alarms and is a prerequisite for assisting people during crowd emergencies.

7.5.2.1 The privacy preserving approach

The goal of our Crowd Detection solution is to provide real-time estimation regarding the crowd density in the target area. We focused on a solution for estimating crowd of people in an indoor scenario, taking into account privacy related issues and deployment costs. Using the sensor fusion approach, we are able to estimate the crowd density by sampling the area with carefully positioned sensors in the indoor environment, which will be used to measure the human activities and correlate them to the density of the crowd present on the scene. In addition, using more traditional sensors such as sound, pressure or CO_2 sensors, inexpensive and privacy preserving infrared proximity sensors is a core part of our approach.

Our solution has the advantage that it estimates the crowd levels with low cost sensors and without infringing any individual's privacy rights. In fact, as Chan et al. describe in [36], there are various Crowd Detection solutions based on computer vision on the market, however privacy is a well-known problem for computer vision technologies for two reasons: first, the perception of compromised privacy is particularly strong for technologies which record the people's actions; second, current vision-based monitoring is usually based on object tracking or image primitives, both of which imply the identification of the individuals.

The NEC Crowd Detection solution has been deployed and tested for a trial in a Singapore shopping mall, where our system has been running for a period of two months and proving the feasibility of such solution. During the trial 23 sensors were deployed using a similar sensor installation plan as described in Figure 7.7 for estimating the crowd levels.

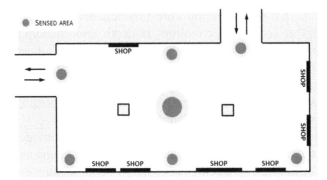

Figure 7.7   Example of sensor installation in a shopping mall.

7.5.3 Mobile Operation Centres (MOC)

The Crowd Detection solution shows how IoT can be used to enhance people safety for indoor environments. However, for big public events like the Football World Cup or the Olympics, IoT analytics is also required to preserve people safety. During such events, many agencies are collaborating to ensure the people safety. As more and more sensors are deployed in urban area by different city service departments like police, fire department, homeland security etc., it is extremely important to share sensor information and derived situations across different agencies in order to improve the safety of such big events.

Typically, a Central Control Centre is designed to deal with normal tasks but has limited amount of resource to handle big events. To overcome such a problem, the capabilities of the control centres can be enhanced by deploying mobile operation centres (MOC), which can be easily setup in a fixed location. Traditionally, MOCs are equipped with voice communication and video cameras to capture critical situations. With the help of the IoT systems, we are able to capture more information about the real world using sensors that have been built into wearables devices, attached to the normal tools of a law enforcement officers, or built into devices like cars, riot control barriers, entrance gates, etc. Still, human intelligence is needed to understand the situations and react to them even with the assistance of derived information from IoT analytics. In this context, ensuring dynamic information flow between the physical world and different authorities is important for enhancing the people's safety.

NEC developed a Mobile Operational Centre (MOC) solution for inter-agency collaboration in a Smart City, which enables the dynamic data exchange of real-time sensor data streams between different agencies. The

MOC is realized combining a dynamic and federated IoT system with an IoT discovery component [38], which is able to handle presence registration of resources with their locations, types etc., and an IoT broker [37], which is able to fetch data by querying/subscribing to the IoT discovery component and requesting data from the underlying data sources. In such system IoT analytics play a role of mining and visualizing the sensor data in real-time to be used within a dashboard shown in Figure 7.8.

7.5.4 Conclusions and Outlook

As sensors and actuators are becoming cheap and being widely deployed in modern cities like Santander in Spain and Chicago in US, the Internet of Things is now providing us great potential to improve our society in terms of safety, security, efficiency and equality by leveraging collected data. Our research goal in IoT analytics is turn collected data into actionable insights to improve and ensure Public Safety in various business domains. For Public Safety, the main challenge to be addressed is to sense critical situations and act on them in real-time. This paper introduced the major technical issues that we are trying to solve in the IoT analytics area, in terms of privacy-preserving, sensor data fusion, anomaly detection, and dynamic data exchange.

Two concrete solutions have been presented in detail to explain how we improve Public Safety using IoT analytics techniques at different levels.

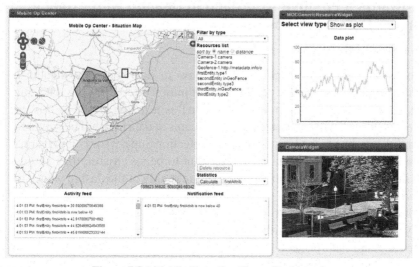

Figure 7.8 Mobile Operation Center Dashboard.

Regarding the Crowd Detection system, additional advantages compared to the existing solutions are: the approach is privacy preserving, the amount of sensors required scales better with the area to be monitored and the cost of the individual sensors and computing nodes is considerably lower than the hardware commonly used in the state-of-the art. In addition to that, our solution estimate crowd levels in real-time and this is also an advantage for trigger quick actions and preserve the people safety without compromise their privacy.

The second is the Mobile Operational Centre solution, enabling dynamic data exchange of real-time sensor data streams between different agencies. This can be used by city authorities like police offices to quickly enable inter-agency collaboration. Both solutions have been deployed and tested in Singapore as part of the Safer City solution.

As future work, we intend to complement the current approaches by addressing additional challenges in the IoT analytics for Public Safety. In the case of Crowd Detection, we are exploring how reinforcement learning can improve the current solution in real-time without losing in system performance or how to distinguish the crowd estimation from other type of emergency without a human interaction. In addition to that, we are extending the Crowd Detection solution to outdoor areas exploring new techniques like Bluetooth and Wi-Fi monitoring.

7.6 Towards a Positive Approach in Dealing with Privacy in IoT Data Analytics

7.6.1 Introduction

Businesses are looking for guidance on how to deal with big data in a responsible/legal way as they see the opportunities offered by big data, big data generation, collection, and analytics. IoT is a major driver in this, as "connected things" will generated endless streams of data that will be captured and used. According to the European Data Protection Supervisor Peter Hustinx (December 2014): "If big data operators want to be successful, they should invest in good privacy and data protection, preferably at the design stages of their projects".

While we want to benefit from the value that IoT and its data have to offer, the key outcome should be "trust" by citizens and consumers. This requires that privacy and data protection are taken into account in every step of the development cycle of IoT technologies and services.

7.6.2 IoT and Privacy

It is clear: in terms of pervasiveness, IoT has already contributed to the emergence of a society in which almost everything is or can be monitored, well beyond the des criptions as used by George Orwell in his book "1984" [42]. The novel is set in Airstrip One (formerly known as Great Britain), a province of the super state Oceania in a world of perpetual war, omnipresent government surveillance, and public manipulation, dictated by a political system euphemistically named English Socialism (or Ingsoc in the government's invented language, Newspeak) under the control of a privileged Inner Party elite that persecutes all individualism and independent thinking as "thought crime". "Big Brother is watching you", and trust in society and freedom is sketched as very low (Figure 7.9). This book had a great influence of the thinking of a generation that grew up after World War 2 and reflects some of the thinking that is fundamental in the discussions about privacy.

Now: whereas the levels of monitoring are very high and well beyond the imagination of Orwell in terms of what technically is possible, in Europe trust in government and society has remained at a relatively high level. When Snowden revealed, starting in June 2013, some evidence reflecting the pervasiveness of monitoring through numerous global surveillance programs, many of them run by the NSA and the Five Eyes[1] with the cooperation of

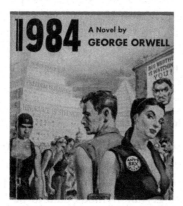

Figure 7.9   1984, a society in which you can trust nobody – and "Big brother" sees it all, and a reality of pervasive monitoring by security forces in 2013 [43].

---

[1]"Five Eyes", often abbreviated as "FVEY", refer to an intelligence alliance comprising Australia, Canada, New Zealand, the United Kingdom, and the United States that was formed. These countries are bound by the multilateral Agreement, a treaty for joint cooperation in signals intelligence.

telecommunication companies and European governments, this resulted in widely expressed concern and even outrage by the general public, civil society and politicians.

This led to a global discussion making clear that monitoring is a necessity, yet should not take place at all costs, and a balance is yet to be found. This results in a discussion that will continue to stretch over the decades to come.

7.6.3 European Way Forward

Within this setting, the discussion in Europe about privacy and data protection is finding its way, moving from a Directive on Data protection and privacy towards an anticipated General Data Protection Regulation. The reform aims to strengthen individual rights and tackle the challenges of globalisation and new technologies. It furthermore attempts to "simplify" compliance as the Regulation would become directly applicable law in all EU member states, whereas the Directive was implemented through national Privacy Acts in similar but not always identical ways.

When the original Data Protection Directive was developed and agreed in 1995, the Internet was by far not as important as today, and nobody had even mentioned the term "Internet of Things" yet. The current reform has been under way since 2011 and culminated in a Proposal to Council and Parliament by the European Commission on 25 January 2012. This proposal was approved by the European Parliament in March 2014, and, assuming that a compromise can be reached in the course of this year, is expected to come into force in 2017. It should be noted however that significant differences still exist between the Commission, Council and Parliament, so that a consensus text in 2015 is not yet a certainty.

7.6.4 Challenges Ahead

Yet, when the Regulation was first being discussed in 2011, "big data" was not yet a widely recognised issue. Today, we know that big data, and big data analytics, fundamentally challenge the concept of "personal data". Big data analytics allows seemingly anonymous data to be linked together and correlated in order to allow individual persons to be identified. A recent Opinion from the Article 29 Working Party – Europe's pre-eminent data protection body – recognises the value of IoT, but also the potential intrusions it can generate to privacy. In this Opinion, statements are made that alarmed businesses around the world, as what is suggested may put a lock on many current developments in the field.

Legal uncertainty remains on many issues, even if it is clear that current law also applies to IoT applications used to collect or analyse personal data. Business are looking for guidance on this, as big data is a subject of interest to many, and companies around the world are looking into the opportunities offered by big data, big data generation, collection, and analytics.

As already noted above, the European Data Protection Supervisor Peter Hustinx has stressed the importance of investment in solid privacy and data protection, and recognizes the role of "soft law" on this point. These investments can drive the innovation, development and deployment of IoT, and are a pre-condition for European (co-)sponsored research.

7.6.5 Way Forward

A way forward could include the habit/obligation of a Privacy Impact Assessment in the design stage of new IoT products and services, and conscious implementation of Privacy Enhancing Technologies and Methods from the outset when thinking of which (and how) data to collect, store, and share. This approach would ensure that new ideas are not hamstrung by regulation, but rather that a culture of privacy awareness and advance consideration is promoted: the impact of any new IoT solution on individuals should be considered prior to deployment, rather than as an issue that may require fixing afterwards.

We need IoT to deal with certain societal challenges. As IoT in combination with big data analytics brings a paradigm shift in ways that data can be related to people, it will take a number of years to come to a better understanding on how to deal with this.

Legislation related to consumer protection, ranging from product safety, to product reliability, product information reliability and personal data protection, tends to be static and oblivious to rapid technology shifts. We need to ensure that the application of the law reflects an understanding of the sensitivity of data in a big data and big data analytics context.

From an innovation and deployment perspective, it will be important to design products and services in such a way that they can continue to serve (local?) society even if values and choices are different in different markets, and/or change over time.

This requires transparency (what data are collected, in what way, how are they stored and unlocked, and who has access to it), accountability (if someone is not using the data in a correct, authorized way – who is accountable for taking action), and choice (can I adapt the settings related to IoT in my environment to my specific legal and cultural preferences?). This goes beyond mere regulatory actions, and requires greater awareness of where sensitive issues may emerge, and the implementation of robust and flexible technological solutions that can be tailored to reflect these changes in society.

7.6.6 Conclusions and Outlook

It is highly important to ensure that our European privacy approach does not prevent the use of data, but rather that it prevents abuse of data – simply because:

- Increasingly almost all data will be relatable to persons, from the outset, and we need to find a way to deal with that responsibly;
- We cannot afford not using data at large scale, both from societal perspective, and as there is clear commercial (thus economic) value;

If Europe wants to benefit from the emerging opportunities arising with IoT – and it is the opinion of the authors that this is a boat Europe cannot afford to miss – we will need to use data in a responsible way (both collecting, storage and sharing – and actively fight abuse).

"Going ethical" when building IoT products and services can bring us new growth and innovation and helps us to create a world we want our children to live in, respecting European values, including privacy but also and perhaps more importantly transparency and choice. The law cannot do it all for us. It is our own standards and ethics that will transform the world. Hence, legislation enabling, supporting and promoting these priorities is required.

Ethical behavior is a cultural thing: it needs to be embraced, lived, in every aspect of the business. It needs to be talked about and to be an explicit value. Ethics is a living thing and can only thrive when welcomed and constantly encouraged.

Acknowledgment

Part of the work described in Section 7.3 *Cloud-Based IoT Big Data Platform* has been carried out in the scope of the Horizon 2020 iKaaS project (iKaaS.com) (Grant Agreement number 643262).

Bibliography

[1] P. Barnaghi, A. Sheth, and C. Henson. "From Data to Actionable Knowledge: Big Data Challenges in the Web of Things", *IEEE Intelligent Systems*, vol. 28, no. 6, pp. 6–11, Nov/Dec. 2013.

[2] J. Gama, I. Zliobaite, A. Bifet, M. Pechenizkiy, A. Bouchachia. "A survey on concept drift adaptation", *ACM Computer Surveys* 46, 4, article 44, 2014.

[3] P. Barnaghi et al. I. Borthwick. (editor), "Digital Technology Adoption in the Smart Built Environment", *IET Sector Technical Briefing, The Institution of Engineering and Technology (IET)*, Technical report, March 2015.

[4] H. Karl and A. Willig. "Protocols and Architectures for Wireless Sensor Networks", Wiley-Blackwell, 2007.

[5] S. Nastic, S. Sehic, H. L. Truong, S. Dustdar. "Provisioning Software-defined IoT Cloud Systems", *The 2nd Int. Conf. on Future Internet of Things and Cloud (FiCloud-2014)*, 2014.

[6] M. Compton et al, "The SSN ontology of the W3C semantic sensor network incubator group", *Web Semantics: Science, Services and Agents on the World Wide Web,* vol. 17, pp. 25–32, 2012.

[7] J. Lin, E. Keogh, S. Lonardi. et al., "A Symbolic Representation of Time Series, with Implications for Streaming Algorithms", *Proceedings of the 8th ACM SIGMOD workshop on Research issues in data mining and knowledge*, 2003, pp. 2–11.

[8] Y. Bengio, A. Courville, P. Vincent. "Representation Learning: A Review and New Perspectives", *IEEE Trans. PAMI, special issue Learning Deep Architectures*, 2013.

[9] P. Barnaghi, A. Sheth, A. Singh, M. Hauswirth. "Physical-Cyber-Social Computing: Looking Back, Looking Forward", Guest Editors Introduction, *IEEE Internet Computing*, May/June, 2015.

[10] Cisco, "Cisco Visual Networking Index: Global Mobile Data Traffic Forecast Update 2014–2019 White Paper," 03-Feb-2015. [Online]. Available: http://www.cisco.com/c/en/us/solutions/collateral/service-provider /visual-networking-index-vni/white_paper_c11-520862.html. [Accessed: 23-May-2015].

[11] HM Government, "Personalised health and care 2020: a framework for action," 13-Nov-2014. [Online]. Available: https://www.gov.uk/gover nment/publications/personalised-health-and-care-2020/using-data-and-t echnology-to-transform-outcomes-for-patients-and-citizens. [Accessed: 23-May-2015].

[12] A. Bassi, M. Bauer, M. Fiedler, T. Kramp, R. van Kranenburg, S. Lange, and S. Meissner, Eds., Enabling things to talk: designing IoT solutions with the IoT architectural reference model; [Iot-A, Internet of things – architecture]. Heidelberg: Springer, 2013.

[13] H. R. Schindler, J. Cave, N. Robinson, V. Horvath, P. Hackett, S. Gunashekar, M. Botterman, S. Forge, and H. Graux, Europe's policy options for a dynamic and trustworthy development of the Internet of Things: SMART 2012/0053. Santa Monica, CA: RAND Corporation, 2013.

[14] EC: European Political Strategy Centre, "Ethics, science and technology," 25-Mar-2015. [Online]. Available: http://ec.europa.eu/epsc/ege_en.htm. [Accessed: 23-May-2015].

[15] E. Niehaus, M. Herselman, and A. N. Babu, "Principles of Neuroempiricism and generalization of network topology for health service delivery," Indian J. Med. Inform., vol. 4, no. 1, 2009.

[16] O. Ferrer-Roca, R. Tous, and R. Milito, "Big and Small Data: The Fog," 2014, pp. 260–261.

[17] C. Thuemmler, J. Mueller, S. Covaci, T. Magedanz, S. de Panfilis, T. Jell, and A. Gavras, "Applying the Software-to-Data Paradigm in Next Generation E-Health Hybrid Clouds," in Proceedings of the 2013 10th International Conference on Information Technology: New Generations, Washington, DC, USA, 2013, pp. 459–463.

[18] S. Nunna, A. Kousarida, M. Ibrahim, M. Dillinger, C. Thuemmler, H. Feussner, and A. Schneider, "Enabling Real-Time Context-Aware Collaboration through 5G and Mobile Edge Computing," in Proceedings of Information Technology New Generations 2015, 2015.

[19] R. Vargheese and H. Dahir, "An IoT/IoE enabled architecture framework for precision on shelf availability: Enhancing proactive shopper experience," 2014, pp. 21–26.

[20] N. Khan, "Fog Computing: Better Solutions for IT," Int. J. Eng. Tech. Res., vol. 3, no. 2, Feb. 2015.

[21] R. Bond, "Big Data and Healthcare," Charles Russell Speachlys. [Online]. Available: http://www.charlesrussellspeechlys.com/updates/publications/commercial-new/big-data-and-healthcare/?UTM_SOURCE=MONDAQ&UTM_MEDIUM=SYNDICATION&UTM_CAMPAIGN=VIEW-ORIGINAL. [Accessed: 23-May-2015].

[22] Deloitte, "Big data revolution: six trends unlocking the power of health care analytics," A view from the Center, 10-Feb-2014. [Online]. Available: http://blogs.deloitte.com/centerforhealthsolutions/2014/02/big-data-revolution-six-trends-unlocking-the-power-of-health-care-analytics.html#.VWB75HuJgb4. [Accessed: 23-May-2015].

[23] C. Tan, L. Sun, and K. Liu, "Big Data Architecture for Pervasive Health-care: A Literature Review," presented at the 23rd European Conference on Information Systems, Muenster, 2015.

[24] MarketsAndMarkets, "Healthcare Mobility Solutions Market by Prod-ucts & Services (Mobile Devices, Mobile Apps, Enterprise Plat-forms), Application (Patient Care, Operations, Workforce Management), End Users (Payers, Providers, Patients) – Global Forecast to 2020," May 2015.

[25] MarketsAndMarkets, "Global Healthcare Cloud Computing Market worth $5.4 Billion by 2017." [Online]. Available: http://www.marketsand markets.com/PressReleases/cloud-computing-healthcare.asp

[26] F. Fernandez and G. C. Pallis. "Opportunities and challenges of the Internet of Things for healthcare: Systems engineering perspective," in Wireless Mobile Communication and Healthcare (Mobihealth), 2014 EAI 4th International Conference on, 2014, pp. 263–266.

[27] A. Banafa, "Fog Computing: From the Center to the Edge of the Cloud," New Trends in Hi Tech, 22-Aug-2014.

[28] M. Ectors, "Fog Computing Might Save Operators From an IoT Data Tsunami," DZone: Smart Content for Tech Professionals, 07-Feb-2014.

[29] Data Protection Working Party, "Opinion 4/2007 on the concept of personal data." 20-Jun-2007.

[30] A. Paulin. "Towards Self-Service Government – A Study on the Com-putability of Legal Eligibilities," J. Univers. Comput. Sci., vol. 19, no. 12, pp. 1761–1791, Jun. 2013.

[31] E. Bertino, G. Ghinita, and A. Kamra. Access control for databases concepts and systems. Boston: Now, 2011.

[32] A. Paulin, "Towards the Foundation for Read-Write Governance of Civilizations," in Third International Conference on Software, Services and Semantic Technologies S3T 2011, vol. 101, D. Dicheva, Z. Markov, and E. Stefanova, Eds. Berlin, Heidelberg: Springer, 2011, pp. 95–102.

[33] A. Paulin and T. Welzer. "A Universal System for Fair Non-Repudiable Certified e-Mail without a Trusted Third Party," Comput. Secur., 2013.

[34] L. Sanchez, R. Ramdhany, A. Gluhak, S. Krco, E. Theodoridis, D. Pfisterer, L. Muñoz; J. A. Galache, P. Sotres, J. R. Santana, V. Gutierrez. SmartSantander: IoT Experimentation over a Smart City Testbed.

[35] NEC public safety whitepaper [Online]: http://www.nec.com/en/global/ solutions/safety/pdf/Safer_Cities_WP.pdf

[36] Chan, Antoni B., Z-SJ Liang, and Nuno Vasconcelos. *"Privacy preserving crowd monitoring: Counting people without people models or tracking."* *IEEE Conference on Computer Vision and Pattern Recognition (CVPR)*, 2008.

[37] NEC IoT Broker component [Online]: https://github.com/Aeronbroker/Aeron

[38] M. Bauer and S. Longo. "Geographic Service Discovery for the Internet of Things." Ubiquitous Computing and Ambient Intelligence. Personalisation and User Adapted Services. Springer International Publishing, 2014. 424–431.

[39] Cisco IoT Forecast [Online]: http://share.cisco.com/internet-ofthings.html

[40] H. Soliman, K. Sudan, A. Mishra. *A smart forest-fire early detection sensory system: Another approach of utilizing wireless sensor and neural networks, IEEE Sensors*, vol., no., pp. 1900, 1904, 1–4 Nov. 2010.

[41] NEC water leak detection service [Online]: http://www.nec.com/en/global]/solutions/waterloss-management/

[42] G. Orwell. *Nineteen Eighty-Four. A novel. London*: Secker & Warburg, 1949.

[43] G. Farvell., http://www.csmonitor.com/Commentary/Monitor-Political-Cartoons, retrieved 10 Jan. 2015.

8

Internet of Things Experimentation: Linked-Data, Sensing-as-a-Service, Ecosystems and IoT Data Stores

Martin Serrano[1], John Soldatos[2], Philippe Cousin[3] and Pedro Maló[4]

[1]National University of Ireland – Insight Centre for Data Analytics, Galway, Ireland
[2]Athens Information Technology, Athens, Greece
[3]e-Global Market – Sophia Antipolis, Nice, France
[4]Faculdade de Ciências e Tecnologia da Universidade Nova de Lisboa, UNINOVA, Lisbon, Portugal

8.1 Introduction

The vision of integrating IoT platforms, testbeds and their associated silo applications is related with several scientific challenges, such as the need to aggregate and ensure the interoperability of data streams stemming from different IoT platforms or testbeds, as well as the need to provide tools and techniques for building applications that horizontally integrate diverse IoT Solutions. The convergence of IoT with cloud computing is a key enabler for this integration and interoperability, since it allows the aggregation of multiple IoT data streams towards the development and deployment of scalable, elastic and reliable applications that are delivered on-demand according to a pay-as-you-go model. During the last years we have witnessed several efforts towards IoT/cloud integration (e.g., [1, 2]), including open source implementations of middleware frameworks for IoT/cloud integration [3, 4] and a wide range of commercial systems (e.g., Xively (xively.com), ThingsWorx (www.thingsworx.com), ThingsSpeak (www.thingspeak.com), Sensor-Cloud (www.sensor-cloud.com)). While these cloud infrastructures provide the means for aggregating data streams and services from multiple

IoT systems, they are not fully sufficient for alleviating the fragmentation of IoT platforms, facilities and testbeds, this is because they emphasize on the syntactic interoperability (i.e. homogenizing data sources and formats) rather on the semantic interoperability of diverse IoT platforms, provided services and data streams.

Advances in the Internet of Things (IoT) area have progressively moved in different directions (i.e. designing technology, deploying the systems into the cloud, increasing the number of inter-connected entities, improving the collection of information in real-time and not less important the security aspects in IoT). IoT Advances have drawn a common big challenge that focuses on the integration of the IoT generated data. The key challenge is to provide a common sharing model or a set of models organizing the information coming from the connected IoT systems, and most important is that the coming generation of IoT technology and systems are prepared to support the common model(s). Solutions to the problem of formulating and managing Internet of Things data from heterogeneous systems and environments (i.e. environments comprising cities, industry, agriculture, etc.) and entity resources (such as smart devices, sensors, actuators, etc.) in cloud infrastructures remains as a key challenge.

Several IoT projects [5] have started work on the semantic interoperability of diverse IoT platforms, services and data streams. To this end, they leverage IoT semantic models (such as the W3C Semantic Sensor Networks (SSN) ontology [6, 7]) as a means of achieving interoperable modelling and semantics of the various IoT platforms. A prominent example is the FP7 OpenIoT project, an awarded by BlackDuck Software Co. winner of the open source project for 2013. OpenIoT has developed and released an open source blueprint infrastructure [8] for the semantic interoperability of diverse sensor networks at a large scale, the source code and the documentation of the project are available at https://github.com/OpenIotOrg/openiot.

The semantic interoperability of diverse sensor clusters and IoT networks is based on the virtualization of sensors in the cloud. At the heart of these virtualization mechanisms is the modelling of heterogeneous sensors and sensor networks according to a common ontology, which serves as harmonization mechanism of their semantics, but also as a mechanism for linking related data streams as part of the linked sensor data vision. This virtualization can accordingly enable the dynamic discovery of resources and their data across different/diverse IoT platforms, thereby enabling the dynamic on-demand formulation of cloud-based IoT services (such as Sensing-as-a-Service services). Relevant semantic interoperability techniques are studied in depth as part of the fourth activity chain of the IERC cluster (IERC-AC4) (see for example [9]). Similar techniques could serve as a basis for unifying and integrating/linking geographically and administratively dispersed IoT testbeds, including those that have been established as part of Future Internet Research and Experimentation platform projects (FIRE). Such integration holds the promise of adding significant value to all of the existing IoT testbeds, through enabling the specification and conduction of large-scale on-demand experiments that involve multiple heterogeneous sensors, Internet Connected Objects (ICOs) and data sources stemming from different IoT testbeds.

In this paper the design principles for IoT Linked-Data with a vision towards IoT data sharing are introduced and the design principles for a framework and set of common models for using the generated IoT information in the context of utility-driven cloud-based Experimentation are described. This paper also presents the main building blocks of a federated environment towards building an Ecosystem of IoT shared data, which emphasizes on-demand establishment of IoT Experimentation services based on the federated formulation of information and data sets. The presented federated framework called FIESTA-IoT leverages well-known IoT technologies and standards in the form of IoT data stores from IoT data testbeds. The structure of the paper is as follows: Section 8.2 presents the main characteristics of Experimentation as a Service in the form of design principles. Section 8.3, delves into more details about Linked Data, Global Information Systems and IoT Infrastructures and the specification of experimentation services to the common models to facilitate common information services. Section 8.4 describes federation in the context of ecosystems and data stores by means of federated IoT services. Section 8.5 describes FIESTA-IoT and the IoT data streams and IoT-experimentation models enabling experimentation as a service (EaaS) by means of available data sets from IoT testbeds and describes some sample applications that will be integrated towards the validation of the federated framework. Finally, Section 8.6 concludes the paper with the most relevant references.

8.2 Experimentation as a Service

Based on the Sensing-as-a-Service paradigm, dynamic virtualized discovery capabilities for IoT resources could give rise to a more general class of Experiment-as-a-Service (EaaS) applications for the IoT domain. EaaS services are executed over converged IoT/cloud platforms, which additional developments are required on the basis of adapting technologies for exchanging data and interacting each other for mutual sharing of available resources. EaaS services are not confined to combinations of sensor queries (such as Sensing-as-a-Service), but they would rather enable the execution of fully-fledged experimental data workflows comprising actuating and configuration actions over the diverse IoT devices and testbeds when offered as services. The benefits resulting from the establishment and implementation of an EaaS paradigm for the IoT domain include:

- The expansion of the scope of the potential applications/experiments that will be designed and executed. Specifically, the integration of diverse testbeds will offer European experimenters/researchers with the possibility of executing IoT experiments that are nowadays not possible. Some possible examples in the areas of pollution management, crisis management and commercial applications are described in [10].
- The ability to repurpose IoT infrastructures, devices and data streams in order to support multiple (rather than a single) applications. This will increase the ROI associated with the investment in the testbeds infrastructure and software. For testbeds built based on EC co-funding, this will also maximize value for EC money (invested on testbeds).
- Possibility for sharing IoT data (stemming from one or more heterogeneous IoT testbeds) across multiple researchers. This can be a valuable asset for setting up and conducting added value IoT experiments, since it will enable researchers to access data in a testbed agnostic way i.e. similar to accessing a conventional large scale IoT database.
- The emergence of opportunities for innovative IoT applications, notably large scale applications that transcend multiple application platforms and domains and which are not nowadays possible.
- The avoidance of vendor lock-in, when it comes to executing IoT services over a provider's infrastructure, given that an EaaS model could boost data and applications portability across diverse testbeds.

Beyond the interconnection and interoperability of IoT testbeds, semantic interoperability tools and techniques could also enable the wider interoperability of IoT platforms, which is a significant step towards a global IoT ecosystem. Figure 8.1 depicts the experimentation as a service scenario where the data sources available for IoT silos are made available to third party experimenters (users) by means of accessibility to the data in programmable way with capacity for resource allocation. The involvement of third parties will therefore play an instrumental role for the large-scale validation of the

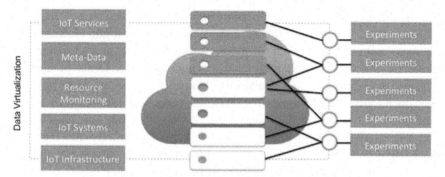

Figure 8.1 An "Experimentation-as-a-Service" Scenario representation.

EaaS infrastructures, but also for the take-up of the global market confidence building IoT interoperability solutions.

8.3 Linked Data, Global Information Systems and IoT Infrastructures

According with the last report from Gartner on emerging technologies and particularly on IoT [11], we are heading towards a world of billions of things, IoT devices are expected to generate enormous amount of (dynamically distributed) data streams, which can no longer be processed in real-time by the traditional centralized solutions. IoT needs a distributed data management infrastructure to deal with heterogeneous data stream sources that autonomously generates data at high rates [12]. An early system designed to envision a worldwide sensor web [13] is IrisNet, which supports distributed XML processing over a worldwide collection of multimedia sensor nodes, and addresses a number of fault-tolerance and resource-sharing issues. A long the same line, HiFi [14] also supports integrated push-based and pull-based queries over a hierarchy where the leaves are the sensor feeds and the internal nodes are arbitrary fusion, aggregator, or cleaning operators.

A series of complementary database approaches aimed to provide low-latency continuous processing of data streams on a distributed infrastructure. The Aurora/Medusa [15], Borealis [16], and TelegraphCQ [17], StreamGlobe [18], StreamCloud [19] are well-known examples of this kind. These engines provide sophisticated fault-tolerance, load-management, revision processing, and federated-operation features for distributed data streams. A significant portion of the stream processing research merit of these systems has already made its way from university prototypes into industry products such as TIBCO Stream Base, IBM Stream Info Sphere, and Microsoft Streamlight. However, such commercial products are out of reach of most IoT stream applications and there have not been any comprehensive evaluation in terms of cost effectiveness, performance and scalability. Due to this reason, there have emerged open source stream processing platforms from Apache Storm [20], S4 [21] and Spark [22] which were primarily built for some ad-hoc applications: Twitter, Yahoo!. While these platforms aim to support elasticity and fault-tolerance, they only offer simple generic stream processing primitives that require significant effort to build scalable stream-based applications.

The above systems provide steps in the right direction for managing IoT data streams in distributed settings. However, they have several federation restrictions in terms of systems of systems and system data organization. For system organization, most of distributed stream processing engines are extended from a centralized stream-processing engine to distributed system architectures. Thus, in order to enable the federation among stream processing sites, they have to follow strictly predefined configurations. However, in IoT settings, heterogeneous data stream sources are provided by autonomous infrastructures operated on different independent entities, which usually do not have any prior knowledge about federation requirements. In particular, a useful continuous federated query might need to compare or combine data from many heterogeneous data stream sources maintained by independent entities. For example, a tourist guide application might need to combine different data stream relevant to the GPS location of users, e.g., weather, bus, train location, flight updates, tourist events. Also, they might then correlate these streams with similar information from other users who have social relationships with the user via social networks such as Twitter, Facebook and also with back ground information like OpenStreetMap, Wikipedia. In such examples, stream data providers did not only agree how their systems will be used to process those federated queries but also they did not agree on data schema/format to make the data accessible via queries for the federated query processing engine.

The need of having uniform and predefined data schema and formats poses various difficulties for query federation on IoT applications using heterogeneous stream data sources. In this sense Linked data for information sharing is widely an accepted best practice to exchange information in an interoperable and reusable fashion way [23], over the Internet different communities use the semantic web standards to enable interoperability and exchange information. Recently linked sensor data has been explored in order to enable the interconnection of multiple new mobile services demand by the rapid development of smart technological devices. Taking a broad view about the state of the art on linked data and its applications in WSN, many of the current problems in WSN still remain in the basic notion of exchanging pieces of information. Data exchanges at sensor level result almost impossible due the capacity for processing information is limited and the stream data processing is very complex.

In global IoT information systems linked sensor data can be exploited in order to facilitate the linking of sensors and ICO related processes at a higher semantic level (within the OpenIoT middleware), where data exchange operations can be conducted. In order to facilitate this process has been necessary to identify the persistent problems as listed below, in order to be addressed and provide alternative solutions.

1. In WSN only a small numbers of services can be offered, which cannot be personalized to meet dynamic wireless sensor configurations.
2. The offered services in WSN are typically technology-driven and static, designed to maximise usage of capabilities of the sensor network rather as individual services and not to satisfy user requirements.
3. The information from sensors cannot be readily adapted to their changing operational context, so sensors cannot communicate their surrounding sensors for configuration purposes, within the objective of changing service usage patterns and rapid new services deployment.

In summary, sensors can only be optimized, on an individual basis, to meet specific low-level objectives, often resulting in sub-optimal operation. By addressing linked data issues, sensor systems can be able to exchange information and thus facilitate customization of sensor networks services. An example of the way how to cope with these requirements and by establishing a WSN design practice is OpenIoT, where it is necessary the composition of data models to generate aggregated data streams. The creation of a data model implies sensor data to be endowed with certain level of information flexibility, operability and management control and most important provide the control platform for interconnected sensors [24]. This every day more popular activity focuses in the semantic enrichment task of the information to generate sensor data models with ontological data to provide an extensible, reusable, common and manageable linked data plane i.e. a plane where aggregate sensor data could be manipulated and operated.

Currently there is a tendency to avoid high-level semantics i.e. ontologies and thus simplified versions of semantics have been generated, i.e. JSON descriptions and its version for linked data JSON-LD. This design approach is more lightweight but as every constraint language, it lacks of enough expressiveness to be used for providing res-usable services and re-programmable application tools. However if well is true JSON-LD is simpler and easy to adopt but not necessarily to adapt into the multiple IoT domains. JSON-LD is focused into the integration of sensor data within the service operations, and offers a more complete understanding of WSN contents based on their continuous data acquisition.

Linked data have demonstrated to be the mechanism to include additional information that facilitates service deployments by means of re-using information, however still lacks in offering re-purposing of technology. Still by using Linked Data in Internet of Things it is expected a more inclusive governance of the management of information and control on devices, likewise facilitates networks (WSN), systems and more adaptive services. Some other major expectations about Linked data in IoT are as follow:

- By linking sensor data information exchange of information within different devices is pursued.
- By supporting the use of data graphs and potentially specialized ontologies in WSN (i.e. W3C SSN) as the mechanism to generate formal descriptions it is possible to represents the collection and formal modelling for sensor networks.
- By using a formal methodology the user's contents represent values used in various service management operations, and thus a knowledge-based approach can be build as part of an inference plane [25].

Semantic annotation aims to be a solution that uses otologies to support interoperability and extensibility required in the systems handling end-user contents for more interoperable applications [27]. Beyond of the formal description of ontology such models need the necessary semantic richness and formalisms to represent different types of information and integrated also service operations.

8.4 Ecosystems and Data Stores by Means of Federated IoT Services

Federation is understood to be: "an organization within which smaller divisions have some internal autonomy" [27]. In the context of the Internet of Things and testbeds, a federation considers that each testbed operates both individually and as part of a larger federation in order to gain value (i.e. larger user base, potential combinations with other testbeds to support richer experimentation, etc.). Represented in Figure 8.2 the typical testbed federation functions are included: resource discovery (finding the required resources for an experiment); resource provisioning (management or resources such that they are available when required); resource monitoring (monitor operation in order to collect experimental results); and finally security (ensuring authorised users can access resources, and the federation provides a trusted base to keep experiment information secure).

Different federation models can then be applied to implement the federation; for example the FedSM project defines a number of models including lightweight federation where there is little if any central control of these functions (by the federation) through to a fully integrated model where a central federation authority implements and provides the functions.

The FIRE programme has a long-standing history in developing cutting edge testbed federations. In the field of networking research: OpenLab provides access to tools and testbeds including PlanetLab Europe, the NITOS wireless testbed, and other federated testbeds to support networking experimentation across heterogeneous facilities. OFELIA is an OpenFlow switching testbed in Europe federating a number of OpenFlow islands supporting research in the Software Defined Networking field. CONFINE co-ordinates

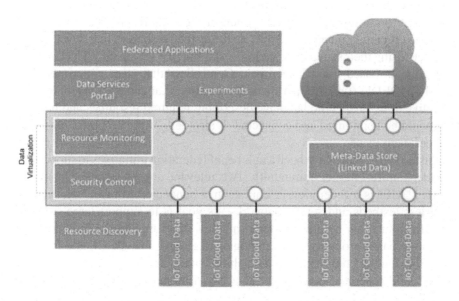

Figure 8.2 Federation Model for the Internet of Things Experimentation.

unified access to a set of real-world community IP networks (wired, wireless, ad-hoc, etc.) to openly allow research into service, protocols and applications across these edge networks. CREW federates five wireless testbeds to support experimentation with advanced spectrum sensing and cognitive radio. Finally, FLEX is a new FIRE project that will provide testbeds for LTE experimentation. In the field of software services, the Bonfire project created a federation of cloud facilities to support experimentation with new cloud technologies. Importantly, in terms of Internet of Things testbeds, SmartSantander provides a set of Smart City facilities through large-scale deployments of sensor networks atop which applications and services can be developed. Also, Sunrise is a federation of sensor network testbeds providing monitoring and exploration of the marine environments and in particular supporting experimentation in terms of the underwater Internet of Things. While each project typically performs federation within its own domain, the Fed4FIRE project is an initiative to bring together heterogeneous facilities across Europe so as to target experimentation across the whole Future Internet field i.e. networks, software, services, and IoT.

Many of the projects (crucially Fed4FIRE) employ OMF [28] and SFA [29] federation technologies. OMF is a control, measurement and management framework for testbeds. From an experimenter's point of view, OMF provides a set of tools to describe and instrument an experiment, execute it and collect its results. From a testbed operator's point of view, OMF provides a set of services to efficiently manage and operate the testbed resources (e.g. resetting nodes, retrieving their status information, installing new OS image). The OMF architecture is based upon Experiment Controllers that steer experiments defined in OEDL (OMF experiment Description Language), which is a declarative domain-specific language describing required resources and how they should be configured and connected. It also defines the orchestration of the experiment itself.

Outside FIRE, there have been a number of federation initiatives to support the wider Future Internet community. Two relevant ones are Helix Nebula and XIFI. XIFI is a federation of data centres connected to resources such as wireless testbeds and sensor networks; its goal is to support large-scale Future Internet trials before transfer to market. XIFI employs a federation architecture based around web technologies (e.g. OAUTH, OCCI, and open Web APIs). On the other hand, Helix Nebula – the Science Cloud is an initiative to build federated cloud services across Europe in order to underpin IT-intense scientific research while also allowing the inclusion of other stakeholders' needs (governments, businesses and citizens).

8.5 FIESTA-IoT: IoT Data Streams and IoT-Experimentation Services

Experimentation as a Service is executed over converged IoT/cloud platforms, Thus additional developments that are required on the basis of adapting technologies for exchanging resources and interacting each other for mutual sharing of available resources. EaaS services are not confined to combinations of sensor queries (such as Sensing-as-a-Service), but they would rather enable the execution of fully-fledged experimental data workflows comprising actuating and configuration actions over the diverse IoT devices and testbeds when offered as services.

The main goal of the FIESTA-IoT project is to open new horizons in the development and deployment of IoT applications and experiments at the EU (and global) scale and based on the interconnection and interoperability of diverse IoT platforms and testbeds. To this end, FIESTA aims for providing a blueprint experimental infrastructure, tools, techniques, processes and best practices enabling IoT testbed/platforms operators to interconnect their IoT Data in an interoperable way, while at the same time facilitating researchers and solution providers in designing and deploying large scale integrated applications (experiments) that transcend the (silo) boundaries of individual IoT platforms or testbeds based o IoT data streams.

FIESTA aims for enabling researchers and experimenters to share and reuse data from diverse IoT testbeds in a seamless and flexible way, which will open up new opportunities in the development and deployment of experiments that exploit data and capabilities from multiple testbeds. Depicted in Figure 8.3, the blueprint experimental infrastructure to be provided by FIESTA will include middleware for semantic interoperability, tools for developing/deploying and managing interoperable applications, processes for ensuring the operation of interoperable applications, as well as best practices for adapting existing IoT facilities to the FIESTA interoperability infrastructure.

In FIESTA we will look at the lack of standards as the major difficulty leading to these restrictions, and the wide (and changing) variety of application requirements. Existing IoT Stream processing engines vary widely in data and query models, APIs, functionality, and optimization capabilities. This has led to some federated queries that can be executed on several IoT stream providers based on their application needs. Semantic Web addresses many of the technical challenges of enabling interoperability among data from different sources. Likewise, Linked Stream Data [30] enables information exchange

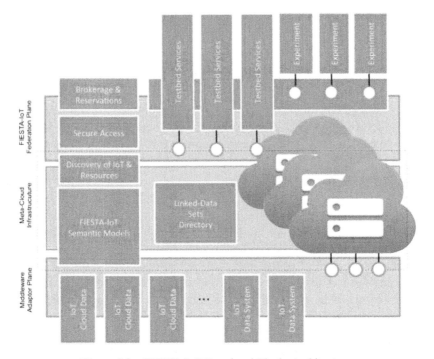

Figure 8.3 FIESTA-IoT Functional Blocks Architecture.

among stream processing entities, i.e, stream providers, stream-processing engines, stream consumer with computer-processable meaning (semantics) of IoT stream data. There have been a lot of efforts towards building stand-alone stream processing engine for Linked Stream Data such as C-SPARQL [31], SPARQL stream [31], CQELS [32], EP-SPARQL [33]. The data and query-processing model of Linked Stream data has been standard by W3C [34]. However, there are only few on-going efforts of building scalable Linked Stream Data processing engines for the cloud like Storm and S4 respectively, i.e., CQELS Cloud [35] and [36]. None of them supports federation among different/autonomous stream data providers.

8.6 FIESTA-IoT for Smart Cities – Semantic Interoperability

A Smart City in conjunction with an ICT expert can use the OpenIoT platform to setup a public cloud infrastructure, which could support multiple (rather than a single) services based on internet-connect objects (e.g., smart phones, sensors). Citizens could then use the provided services through submitting requests to the OpenIoT infrastructure and accordingly having them executed by the middleware. Requests will target information that could be provided on the basis of urban sensors, such as meteorological information, traffic information, parking occupancy information, in conjunction with city maps. OpenIoT will offer the ability to automatically define and formulate such requests, while at the same time providing services in a utility-based fashion.

FIESTA-IoT works on semantic interoperability for IoT and Cloud resources, FIESTA-IoT focuses on developing common annotation models for describing the resources and IoT data and providing validation and testing tools for semantic interoperability evaluation. The core models will be constructed by investigating the existing IoT architectural approaches from OMA Alliance, OneM2M, IPSO Alliance and the IoT-A information models (i.e. resources, service and entity models developed in the FP7 EU IoT-A project, http://epubs.surrey.ac.uk/127271/), and the semantic and ontology models including W3C Semantic Sensor Network Ontology (SSN Ontology) (http://www.w3.org/2005/Incubator/ssn/XGR-ssn-20110628/), EF7 IoT.est and OpenIoT Ontology models. In FIESTA-IoT existing concepts, namespaces and semantic models will be used and a set of core models to describe IoT resources (e.g. sensor devices, gateways, actuators) and their capabilities and features and will also provide semantic models to describe Cloud services and Cloud based components will be adopted and/or created. The existing semantic models such as W3 SSN, IoT-A models are usually developed for specific purposes and in the domain of the projects. These semantic models are also often complex and come with the size and computation overhead especially for annotating large-scale IoT/Cloud resources [37]. Variety of models and annotation schemes also cause the heterogeneity among the semantic descriptions while different namespaces and concepts are used to describe similar resources [38].

8.7 Conclusions

The model of Experimentation as a Service is a new paradigm that follows the model where services are executed over converged IoT/cloud platforms and where additional developments are required on the basis of adapting technologies for exchanging data and interacting each other for mutual sharing of available resources.

EaaS services are not confined to combinations of sensor queries (such as Sensing-as-a-Service), rather they enable the execution of fully-fledged experimental data workflows comprising actuating and configuration actions over the diverse IoT devices and testbeds when offered as services.

The challenges for IoT Experimentation and the main functional blocks for the FIESTA-IoT federated architecture have been presented in this paper and as reference for further developments in IoT federation systems.

The FIESTA-IoT federated architecture has as main objective providing support in the convergence of internet-of-things experimentation platforms by means of federation and facilitating IoT Cloud data in the form of streams for IoT Cloud data experimentation.

Acknowledgements

Part of this work has been carried out in the scope of the project H2020-FIESTA-IoT Project (Federated Interoperable Semantic IoT/Cloud Testbeds and Applications), which is co-funded by the European Commission under Horizons 2020, contract number H2020-ICT-2014-1-643943-FIESTA.

Bibliography

[1] H.M. Mehedi and B. Song. Eui-nam Huh: A framework of sensor-cloud integration opportunities and challenges. *In Proceedings of ICUIMC*. 618–626. 2009.

[2] K. Lee. Extending Sensor Networks into the Cloud using Amazon Web Services, *In Proceedings of IEEE International Conference on Networked Embedded Systems for Enterprise Applications 2010*, 25th November, 2010.

[3] C.G. Fox, S. Kamburugamuve and R. Hartman. Architecture and Measured Characteristics of a Cloud Based Internet of Things API, *In Proceedings of Workshop 13-IoT Internet of Things, Machine to Machine and Smart Services Applications* (IoT 2012) at The 2012 International Conference on Collaboration Technologies and Systems (CTS 2012) May 21–25, 2012 Denver, Co, USA.

[4] J. Soldatos, M. Serrano and M. Hauswirth. Convergence of Utility Computing with the Internet-of-Things, *In proceedings of International Workshop on Extending Seamlessly to the Internet of Things (esIoT)*, collocated at the IMIS-2012 International Conference, 4th6th July, 2012, Palermo, Italy.

[5] M. Bauer, P. Chartier, K. Moessner, N.S. Cosmin, C. Pastrone, J.X. Parreira and R. Rees. *IERC-AC2-Deliverable 1*. 2011.

[6] K. Taylor, Semantic Sensor Networks: The W3C SSN-XG Ontology and How to Semantically Enable Real Time Sensor Feeds, *In proceedings of 2011 Semantic Technology Conference*, June 5–9, San Francisco CA, USA.

[7] M. Compton, P. Barnaghi, L. Bermudez, R.G. Castro, O. Corcho, S. Cox, et. al. The SSN Ontology of the Semantic Sensor Networks Incubator Group, *Journal of Web Semantics: Science, Services and Agents on the World Wide Web*, Elsevier, 2012.

[8] M. Serrano, H. Nguyen, D. LePhuoc, M. Hauswirth, J. Soldatos, N. Kefalakis, P. Jayaraman and A. Zaslavsky. Defining the stack for service delivery models and interoperability in the Internet of Things: A practical case with OpenIoT-VDK. *IEEE Journal on Selected Areas in Telecommunications* 2014.

[9] P. Cousin, M. Serrano and J. Soldatos. Internet of Things Research on Semantic Interoperability to address Manufacturing Challenges, *In Proceedings of the 7th International Conference on Interoperability for Enterprise Systems and Applications, I-ESA 2014*, Albi, France, March, 24–28, 2014.

[10] S. Kirkpatrick, M. Boniface et. al. Future Internet Research and Experimentation: Vision and Scenarios 2020, *AmpliFIRE Support Action (Grant Agreement: 318550) Deliverable D1.1*, June 2013.

[11] Forecast: The Internet of Things, Worldwide, 2014. *Gartner report http://www.gartner.com/document/2625419?* 2014. sthkw=G00259115

[12] M. Balazinska, A. Deshpande, M.J. Franklin, P.B. Gibbons, J. Gray, M. Hansen, M. Liebhold, S. Nath, A. Szalay, and V. Tao. 2007. Data Management in the Worldwide Sensor Web. *IEEE Pervasive Computing* 6, 2 (April 2007), 30–40.

[13] J. Campbell, P.B. Gibbons, S. Nath, P. Pillai, S. Seshan, and R. Sukthankar. 2005 IrisNet: an Internet-scale architecture for multimedia sensors. *In Proceedings of the 13th annual ACM international conference on Multimedia (MULTIMEDIA '05)*. ACM, New York, NY, USA, 81–88.

[14] M.J. Franklin, S.R. Jeffery, S. Krishnamurthy, F. Reiss, S. Rizvi, E. Wu, O. Cooper and A. Edakkunni. Wei Hong: Design Considerations for High Fan-In Systems: The HiFi Approach. CIDR 2005: 290–304.

[15] M. Cherniack, H. Balakrishnan, M. Balazinska, D. Carney, U. Cetintemel, Y. Xing, and S. Zdonik. Scalable Distributed Stream Processing, *In proceedings of 1st Biennial Conf. Innovative Data Systems Research (CIDR)*, 257–268. 2003.

[16] D.J. Abadi, Y. Ahmad, M. Balazinska, U. Cetintemel, M. Cherniack, J.H. Hwang, W. Lindner, A.S. Maskey, A. Rasin, E. Ryvkina, N. Tatbul, Y. Xing, and S. Zdonik. The Design of the Borealis Stream Processing Engine, *In proceedings of 2nd Biennial Conf. Innovative Data Systems Research (CIDR 05)*, 2005. http://www-db.cs.wisc.edu/cidr

[17] S. Chandrasekaran, O. Cooper, A. Deshpande, M.J. Franklin, J.M. Hellerstein, W. Hong, S. Krishnamurthy, S.R. Madden, F. Reiss, and M.A. Shah. 2003. TelegraphCQ: continuous dataflow processing. *In Proceedings of the 2003 ACM SIGMOD international conference on Management of data (SIGMOD '03)*. ACM, New York, NY, USA, 668–668.

[18] B. Stegmaier, R. Kuntschke, and A. Kemper. StreamGlobe: adaptive query processing and optimization in streaming P2P environments. *In Proceedings of the 1st international workshop on Data management for sensor networks* in conjunction with VLDB 2004 (DMSN '04). ACM, NY, USA, 88–97. 2004.

[19] V. Gulisano, R. Jimenez-Peris, M.P. Martinez, C. Soriente, and P. Valduriez. StreamCloud: An Elastic and Scalable Data Streaming System. *IEEE Transactions of Parallel Distributed Systems*, 12 (December 2012), 2351–2365.

[20] Storm: Distributed and fault-tolerant realtime computation. http://storm.incubator.apache.org/

[21] S4: Distributed Stream Computing Platform. http://incubator.apache.org/s4/

[22] Spark: Lightning-fast cluster computing. http://spark.apache.org/

[23] L. Kalinichenko, M. Missikoff, F. Schiappelli, N. Skvortsov. Ontological Modeling *In proceedings of the 5th Russian Conference on Digital libraries*, St. Petesburgo, Russia, 2003.

[24] M. Hauswirth, Manfred, D. Pfisterer and S. Decker, Making Internet-Connected Objects readily useful". *In Interconnecting Smart Objects with the Internet Workshop*, Prague, 2011.

[25] J.M. Serrano, J.C. Strassner and M. ÓFoghlú, A Formal Approach for the Inference Plane Supporting Integrated Management Tasks in the Future Internet *1st IFIP/IEEE ManFI International Workshop*, In conjunction with 11th IFIP/IEEE IM2009, 1–5 June 2009, at Long Island, NY, USA.

[26] M. Serrano, M. Hauswirth and J. Soldatos, Design Principles for Utility-Driven Services and Cloud-Based Computing Modelling for the Internet of Things, *International Journal of Web and Grid Services*, 2014.

[27] OED, *federation n.* OED Online. March 2014. Oxford University Press. http://www.oed.com/view/Entry/68930?redirectedFrom=federation. Access on (March 28, 2014).

[28] T. Rakotoarivelo, M. Ott, G. Jourjon and I. Seskar. OMF: a control and management framework for networking testbeds, *in proceedings of ACM SIGOPS Operating Systems Review* 43 (4), 54–59, Jan. 2010.

[29] J. Augé and T. Friedman (2012) *The Open Slice-based Facility Architecture* available at http://opensfa.info/doc/opensfa.html

[30] D. Le-Phuoc, M. Dao-Tran, M.D. Pham, P. Boncz, T. Eiter, and M. Fink. Linked Stream Data Processing Engines: Facts and Figures. *In proceedings of the Semantic Web–ISWC 2012*, 300–312. 2012.

[31] D.F. Barbieri, D. Braga, S. Ceri, and M. Grossniklaus. An execution environment for C-SPARQL queries. *In Proceedings of the 13th International Conference on Extending Database Technology (EDBT '10)*. ACM, New York, NY, USA, 441–452. DOI=10.1145/1739041.1739095.

[32] D. Le-Phuoc, M. Dao-Tran, J.X. Parreira and M. Hauswirth. A native and adaptive approach for unified processing of linked streams and linked data. *The Semantic Web–ISWC 2011*, 370–388. 2011.

[33] D. Anicic, P. Fodor, S. Rudolph, and N. Stojanovic. EP-SPARQL: a unified language for event processing and stream reasoning. *In Proceedings of the 20th International World-Wide Web Conference*, pages 635–644. ACM, 2011.

[34] RDF Stream Processing Community Group, (online access may 2015) http://www.w3.org/community/rsp/

[35] D. Le-Phuoc, H.N. Mau-Quoc, C. Le-Van and M. Hauswirth. Elastic and Scalable Processing of Linked Stream Data in the Cloud. *In proceedings of International Semantic Web Conference* (1) 2013: 280–297.

[36] J. Hoeksema and S. Kotoulas. High-performance Distributed Stream Reasoning using S4, *In proceedings of Ordering workshop of ISWC2011*. 2011.

[37] P. Barnaghi, W. Wang, C. Henson, K. Taylor, Semantics for the Internet of Things: early progress and back to the future", *International Journal on Semantic Web and Information Systems (special issue on sensor networks, Internet of Things and smart devices)*, September 2012.

[38] P. Barnaghi, P. Cousin, P. Malo, M. Serrano and C. Viho, Simpler IoT word(s) of tomorrow, more interoperability challenges to cope today, *Book chapter in 2013 European Research Cluster for the Internet of Things Book, Internet of Things Converging Technologies for Smart Environments and Integrated Ecosystems*, River Publishers, 2013.

9

Driving Innovation through the Internet of Things – Disruptive Technology Trends

Ovidiu Vermesan[1], Mario Diaz Nava[2] and Hanne Grindvoll[1]

[1]SINTEF, Norway
[2]STMicroelectronics, France

"Chance favors the connected mind." *Steven Johnson*

9.1 Introduction

The development of enabling technologies such as nanoelectronics, communications technologies, sensors/actuators, embedded systems, cloud networking, network virtualization and software will be essential to provide things (medical equipment, vehicles, home appliances, equipment for health- and well-being, monitoring of e.g. assets, pollution, water, wildlife, just to mention a few examples) with the capability to be connected at anytime, anywhere. This development provides a bright future for Internet of Things (IoT) and Industrial Internet of Things (IIoT) product innovations that can influence many different sectors (consumer, industrial, business, etc.). Some of these technologies, such as embedded or cyber-physical systems (CPS), see Figure 9.1, bridge the gap between cyber space and the physical world of real things, and they are crucial in enabling the IoT to deliver its vision and become part of bigger systems in a world of "systems of systems".

Electronic Components and Systems are a pervasive key enabling technology for IoT and CPS, impacting all industrial branches. The cyber-physical systems are defined as "the next generation embedded intelligent ICT systems that are interconnected, interdependent, collaborative, autonomous, and provide computing and communication, monitoring/control of physical components/processes in various applications."

Future CPS need to be scalable, distributed, and decentralized allowing interaction with humans, environment and machines while being connected

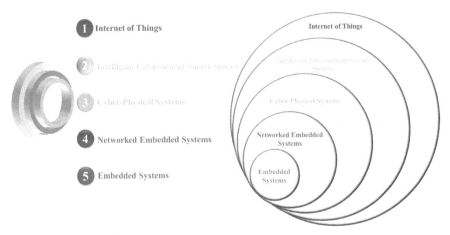

Figure 9.1 Complexity of Systems.

to the Internet or other networks. Adaptability, reactivity, optimality and security are features to be embedded in such systems, as CPS are now forming an invisible 'neural network' of society. CPS leverage cloud computing and IoT capabilities in embedded system's context, while focusing on the various system functions including smart systems integration and a "system of systems" convergence. Cyber-physical systems are one of the most complex systems that form IoT applications.

IIoT adds tougher requirements to the networks for latency, determinism, and bandwidth, while bringing increased complexity considering thousands of connected complex industrial systems that communicate and coordinate the data analytics and actions to improve performance and efficiency and reduce or eliminate downtime. In addition IIoT systems must be adaptive and scalable through software or added functionality that integrates with the overall application solution. In this context, governments will need to make an even bigger investment in digital infrastructure if they are to facilitate the IIoT, as its success is heavily dependent on the presence of robust infrastructures, such as ubiquitous broadband connectivity and sensors.

9.2 Intelligent Edge and Web-Enabled Devices

The IoT development will be realized by creating plug and play, web-enabled, sensing/actuating/communicating devices, by providing common hardware/software/cloud platforms on which they can communicate and exchange information, and by developing new tools and applications.

Enablers are the intelligent edge devices (sensor/actuators, multi com-munication protocols integrated with processing units/microcontrollers and having features like low-power, low cost, embedded security) and the devel-oped hardware/embedded software platforms including development tools.

The different embodiments of the Internet such as IoT and IIoT will be followed by the emergence of the Tactile Internet which is another paradigm shift, in which responsive, reliable network connectivity will enable it to deliver physical, tactile experiences remotely. The requirements will be very high reliability, very low latency and short end-to-end delays (milliseconds) and high network capacity to allow large numbers of devices to communicate with each other simultaneously and autonomously.

For the deployment of IoT technology and applications there is a need for infrastructure development in order to accelerate the IoT adoption. Advancements are needed for integrated sensors/actuators with commu-nication capabilities, sensor interfaces, sensor-specific micro controllers, data management, communication protocols and targeted application tools, platforms and interfaces.

Ubiquitous connectivity is emerging and standards for IoT are used in many applications. Connectivity components, networking hardware, gateways will continue to be profitable and considering the various requirements will be able to differentiate themselves and make money with toolsets for deployment and development, analogous to the offerings of companies in the open source space.

Cloud/Fog computing enables the connection with the edge and provides devices with all the intelligence processed in the cloud. The data captured from the edge devices can be processed and data analytics and data mining algorithms will generate the "smart data" needed for various IoT applications.

The semiconductor stakeholders that focus primarily on the intelligent edge and Web-enabled devices are supporting the development of broader IoT ecosystems and identify roles as both enablers and creators of value for their customers and their customers' customers. This creates partnerships with stakeholders further downstream, such as companies that are building and providing gateways, connectivity, middleware, cloud-based products, services and IoT applications.

A challenge for these IoT ecosystems is that different industries are at different levels of maturity and complexity with respect to the Internet of Things, so the roles that components manufacturers can have in application development in certain industries will vary, as will the timing of growth oppor-tunities. In this context few markets (building/home automation, lighting) have established a number of common APIs (Application Programming Interfaces) and the open question remains related to competing standards. Other markets (monitoring and control systems in manufacturing, IoT technologies in retail) are fragmented and will require longer time to develop. In the case of markets where the standards are under development or missing, semiconductor companies are forming alliances with hardware companies, systems players, and customers to support and assist the development of standards. The smart manufacturing IoT applications are much more complex with most of the hardware platforms that are still proprietary and the data is not accessible outside the close system. In these cases the challenge is to create common standards and keep the compatibility and interoperability with legacy systems. Industry 4.0 and the Industrial Internet Consortium initiatives are aiming to develop such standards in order to fill the gap. The semiconductor stakeholders designing intelligent edge and Web-enabled devices are involved in alliances and standard-setting activities in order to play a role in defining best practices in IoT privacy, security, and authentication, issues that are critical in markets, such as healthcare and wearables, which are dealing with sensitive consumer data.

In this context the strategy for IoT applications is to support a hardware platform by designing a family of devices that are sufficiently flexible to address the needs of multiple industries, that can be used in industrial IoT and consumer IoT applications that have similar characteristics. These devices are covering the application requirements, at one end, high-power, high-performance, application-processing IoT devices, and, at the other end, low-cost, ultralow-power integrated sensors that support sufficient (optimised) functionality and autonomous device operation. These requirements push the semiconductor companies to come with new approaches for product and application development in order to achieve a higher level of design flexibility and to address the IoT applications opportunity.

Many industry research firms, including ABI, Gartner, IDC and IHS, forecast the number of connected devices to be in the range of 20 to 30 billion by 2020. Looking at the semiconductor industry, alone, the growth potential is huge. IoT-related semiconductor sales is expected to grow 19 percent, reaching $5.6 billion in 2015, according to the 2015 IC Insights IC Market Drivers report [5]. The market research firm projects the market will reach $11.5 billion in 2018, see Figure 9.2, translating into a compound annual growth rate of 24.3 percent over the forecast period of 2013 to 2018. The largest IoT semiconductor market segment will continue to be connected cities, including "smart" electric grids, roads and streetlights, and other public infrastructure

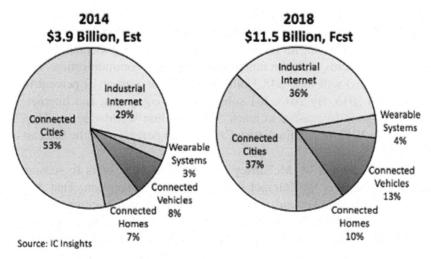

Figure 9.2 IoT Semiconductor sales by systems segments.

applications [4]. Sales for this segment is projected to reach $4.2 billion with a CAGR of 15 percent over the forecast period. Industrial Internet follows with a projected $4.1 billion in sales. Industrial Internet is expected to nearly catch up with connected cities primarily due to high growth in factories, logistics, and medical systems applications.

Semiconductor revenues for the connected homes is expected to surpass the $1 billion mark in 2018 with a CAGR of 32.8 percent, up from $275 million in 2013. Connected automotive systems-mainly in passenger vehicles-offers a high growth potential between 2013 and 2018 with a CAGR of 43.8 percent, reaching $1.5 billion in 2018. Semiconductor sales for wearable systems that connect to the Internet are forecasted to reach $528 million in 2018 with a CAGR of 46.9 percent over the forecast period.

In 2018, IoT-related ICs will account for about three percent of the total $348.1 billion IC market. Beyond embedded IoT subsystems in connected applications, the proliferation of the Internet of Things will expand the use of cloud computing and web servers as well as require upgrades to the overall Internet infrastructure in order to handle growing amounts of data coming from attached systems and things by 2020 [5].

IC Insights forecasts that web-connected things will account for 85 percent of nearly 29.5 billion Internet connections worldwide by 2020 as long as "missing IoT standards" are developed over the next several years, adding to the more than half-dozen existing initiatives for IoT around availability, ease of connection, and compatibility across platforms in different industry sectors. That is up from about 74 percent of the 7.7 billion Internet connections to things in 2010. IC Insights estimates that sales generated by the IoT portion of systems (i.e., the functions for Internet communications and sensor subsystems) will total $48.3 billion in 2014 and grow 19 percent to $57.7 billion in 2015. By 2018, IoT subsystems in equipment and Internet-connected things is forecasted to reach $103.6 billion worldwide, up from $39.8 billion in 2013, translating into a CAGR of 21 percent over the forecast period [4].

The IoT, as defined by McKinsey and the GSA [6], refers to sensors that communicate over the Internet without human intervention. That definition includes wearables and smartphones, both of which communicate autonomously if allowed. It also includes smart devices for vehicles, homes, and cities, and for healthcare and industrial equipment. The report identifies several challenges that can delay the IoT deployment and the McKinsey/GSA survey sees six significant obstacles [6].

Security and privacy of user data. Security is an important requirement for growth in IoT applications. The report noted that the real challenge lies in using available technology to implement end-to-end security solutions for the entire IoT stack-cloud, servers, and devices.

Difficulty building customer demand in a fragmented market. Demand for IoT applications is in its early stages, and its future growth is expected to result from a string of attractive but small opportunities that use a common hardware and software platform, rather than a single "killer application" Semiconductor companies can indirectly play a role in boosting consumer demand by helping developers create innovative applications or by providing assistance to businesses that want to use IoT products and services, including non-traditional clients like start-ups and businesses outside the technology sector-for instance, retailers or hospitals. Users of IoT devices need to be able to derive real, repeatable benefits and ROI from the technology and new business models have to be proposed.

Lack of consistent standards. Some levels of the IoT stack have widely accepted, well-defined standards, but others have none. In still others, there are multiple, competing standards with no obvious winner. Figure 9.3 shows the multiple standards for connectivity. Given the current uncertainty, semiconductor players should pursue a hedging strategy – in other words, focusing on

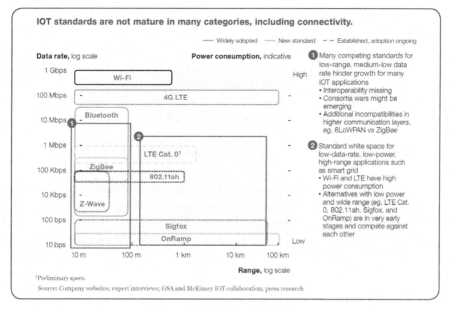

Figure 9.3 IoT Standards [6].

selected standards that are likely to gain widespread acceptance but planning for alternative scenarios. In parallel, semiconductor companies should actively engage with industry associations or other groups that are trying to develop IoT standards, with the goal of supporting the best ones. Such collaboration is important even when companies are trying to help create marketplace standards. This lack of standards is particularly the case among emerging protocols with low-power consumption like LTE Cat.0, 802.11ah, Sigfox, and OnRamp.

A fragmented marketplace with many niche products. Most IoT applications do not generate enough sales to justify design of a single chip specifically targeted at them. The fragmentation represents a concern because it limits economies of scale, thus raising production costs.

Semiconductor companies may be able to achieve the necessary sales volume by classifying IoT devices into archetypes based on their specifications. The companies can then create a single platform to cover each archetype, which will have more widespread appeal than a chip tailored to a niche application.

The challenge of extracting more value from each application. Many semiconductor companies would not extract full value from the IoT if they focused solely on silicon, so they were determined to deliver complete solutions that cover multiple layers of the technology stack. The opportunities include software, security, systems integration and many companies are looking to add value at multiple layers of the technology stack, through software, security, and systems integration, as can be seen in Figure 9.4.

Technological issues that affect the IoT's functionality. Many companies look very confident that they solved the technological needed for the success of the IoT. However, there are companies that consider that technical issues can be a major challenge. Lower power consumption is mentioned as the key improvement that could substantially stimulate demand.

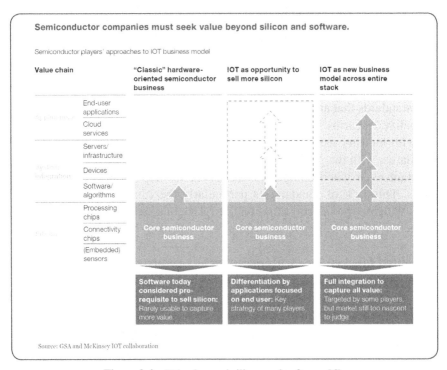

Figure 9.4   Value beyond silicon and software [6].

9.3 IoT Ecosystems

The IoT technology developments have resulted in the creation of many interfaces, devices, operating systems, and applications while the data generated and the security have to be integrated and managed. In this context the success of IoT applications and deployments depends on the involvement of various stakeholders in the value chain that form an ecosystem around various technology platforms. The ecosystems provide efficiencies of scale to develop for customers. In this context the further IoT developments require the use of platform-based approaches that includes flexible hardware architecture deployed across many applications by removing significant amount of the hardware complexity and transform each new problem primarily into a communication and software defined challenge. The same principle needs to be applied to software tools to form powerful hardware-communication-software-analytics platform that creates a unified solution. An effective platform-based approach does not focus on one technology or hardware/software but instead on the innovation within the application itself.

9.4 IoT Platforms

IoT platforms are the highest, most generalized layer of intelligence and user interface, that ties together connected devices and web-based services. They collectively define a reference architecture model for the IoT, taking into consideration a wide range of technologies, communication protocols and standards.

An IoT platform must allow external users and devices to connect to it, based on a governance model, which is the basis to decide "who gets what". IoT platforms must be able to coordinate and manage connectivity issues, and to guarantee the security and privacy of the data exchanged, by a large number of networked devices while overcoming interoperability issues. Agreeing upon a pre-defined set of protocols to share certain services, a federation of platforms will allow optimizing the use of the resources, improving service quality and most likely reducing costs.

Architectural decisions to define IoT platforms must ensure that the developed solution implements a horizontal approach to overcome the existent vertical fragmentation. Furthermore, IoT platforms should address both technological and semantic interoperability issues among heterogeneous IoT devices. Additionally, platforms should be able to minimize the complexity of collecting and processing massive amounts of data generated in IoT scenarios; address scalability and security issues; and guarantee that the developed solution is built upon open-source software, to allow portability and reduce product development costs -while encouraging creativity and collaboration among the various IoT stakeholders.

IoT platforms need to provide solutions to assimilate data from multiple vendors and support open API interfaces across platforms. This requires taking into consideration issues such as: openness, participation, accountability, effectiveness, coherence, etc., while offering innovative solutions that enable self-governance, self-management, and context aware scalability. Following this approach, IoT platforms should include the following elements:

- Abstraction Layer – abstracting physical IoT devices and resources into virtual entities and representations, enabling interoperability through uniform access to heterogeneous devices and resources over multiple communication protocols such as MQTT, Restful, etc.
- Virtualization Layer – providing service look-up mechanisms that bridge physical network boundaries and offer a set of consumable services.
- Data management Framework – enabling storage, caching and querying of collected data as well as data fusion and event management, while considering scalability aspects.
- Semantic Representation Framework – for modelling and management of semantic knowledge.
- Security and Policy Framework – implementing Access Control mechanisms and Federation Identity management responsible for authentication and authorization policies and for enabling federation among several IoT platforms respectively.
- Networking Framework – enabling communication within and across platforms, providing means for self-management (configuration, healing and optimization) through cognitive algorithms.
- Open Interfaces – set of open APIs (possibly cloud-based) to support IoT applications, and ease platform extension by enabling easy interaction and quick development of tools on top of the platform.
- Data Analytics services – providing "real time" event processing, a self-service rule engine to allow users to define simple and complex rules, and querying, reporting and data visualization capabilities.
- Machine learning data analytics – a set of complex machine learning algorithms, for providing real-time decision capabilities.

- Development tools and standardized toolkits – for fast development of (possibly cloud-based) IoT applications that can be integrated by different companies.

Developments of IoT platforms involves an entire ecosystem of stakeholders covering the whole value chain of the IoT that together coordinate and deliver the functionalities and the services required by the various supported IoT applications.

The lack of skills at the application level to capitalize on the IoT, requires a loosely coupled, modular software environment based on APIs to enable endpoint data collection and interaction. The Web platforms using APIs can be used to simplify programming, deliver event-driven processes in real time, provide a common set of patterns and abstractions and enable scale. In this context new tools, search engines and APIs are needed to facilitate rapid prototyping and development of IoT applications. Figure 9.5 shows an example of an IoT device platform developed by ARM.

As IoT applications are being developed and more devices and things are gaining connectivity, platforms for new businesses are expected to emerge in the field of IoT. Following this approach, recent interest areas for emerging platforms include sectors where connectivity is coming in such as smart cities, health care, education and electricity grids sectors. The biggest challenge in these areas concern the governance models, which in its majority are still missing. Additionally, private and public companies from these sectors lack motivation to share already available data and thus it is not possible to extract any value from it. To overcome this, building economic incentives around the openness of data becomes of main importance.

The traditional value chain of the technology sector, where technology companies primarily sold to each other, is not a valid approach for the IoT

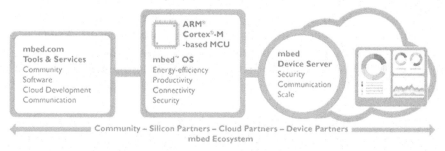

Figure 9.5 ARM mbed IoT Device Platform [10].

ecosystem. The deployment of new dynamic networks in the IoT domain is needed, where connected products and people are in charge of driving the new information values expected from intelligent device networking. IoT ecosystems will be driven by companies able to identify and –innovatively-eliminate barriers to adoption, rather than creating solutions for use cases found within the "empty" spaces.

IoT is a disruptive technology that attracted the interest of many companies in the last few years and as result there are many IoT platforms available for developers to build applications and interconnect devices. An example is the Intel IoT Platform shown in Figure 9.6. These IoT platforms provide hardware components, communication/transport layer, ability to host data in the cloud and the ability to analyze the generated data. The IoT platforms available are platforms that are backed by large companies and consortia such as Allseen Alliance, Homekit, Open Interconnect Consortium, and IBM Internet of Things Ecosystem, ARM, Intel or niche platforms either vertical or use case specific: Litmus automation, Xively, Electric imp, Ayla networks etc. The nature of these platforms has contributed to fragmentation of IoT and IIoT infrastructure and contributed to a race between these companies to own the vertical stack rather than develop products or services that benefit all users and developers. There are few open source platforms in the Internet of Things that can reduce product development costs and encourage creativity and collaboration. Today IoT devices in many cases are installed in their platform and ecosystem.

An overview of existing commercial IoT platforms is given in the following paragraphs.

Axeda [1] – Axeda provides a cloud-based platform for managing connected products and machines and implementing IoT and M2M (machine-to-machine) applications. The platform is used to transform machine data into valuable insights, build and run applications and integrate machine data with other applications and systems to optimize business processes. Axeda's platform encompasses the area of developing and deploying applications and integrating M2M learning into business processes, from preventative data security measures all the way to device provisioning and configuration.

ThingWorx [2] – ThingWorx facilitates the streamlined creation of end-to-end smart applications for agriculture, cities, grid, water, building and telematics. Traditional industries are transformed and equipped with modern-day connectivity and smarter solutions through connected devices that provide

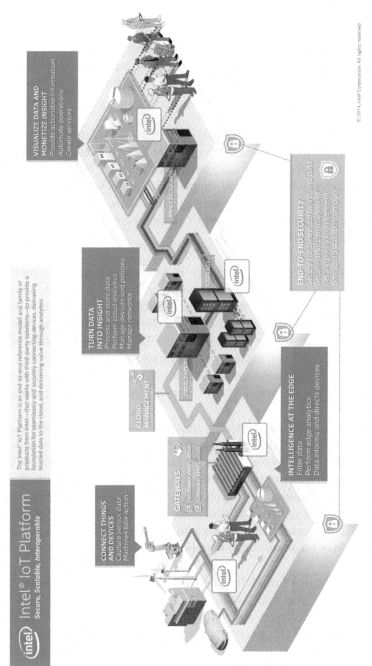

Figure 9.6 Intel® IoT Platform [11].

comprehensive data collection and analysis for data-driven decision-making. ThingWorx reduces the time, cost and risks of building M2M and IoT applications. Users can build comprehensive mobile interfaces with zero coding, take advantage of ThingWorx Composer for application modeling, as well as real-time dashboards and collaborative workspaces – all with the scalability to support millions of devices.

SAP Internet of Things Solutions [3] – SAP's IoT solutions facilitate connectivity and multi-directional communication to enable users to interact with their devices in new ways. Transforming operations in field service and remote asset management, providing supply chain visibility and predicting and remedying logistics bottlenecks are just some of the challenges solved by SAP's remote maintenance and service, connected logistics and connected retail solutions for the IoT.

Microsoft Azure IoT Suite [7] – Azure is Microsoft's cloud-computing platform that provides the infrastructure to build and manage apps in the cloud. Azure Internet of Things Suite is an integrated service that makes use of all the relevant Azure capabilities to connect devices. The suite captures the varied data these devices generate. The Azure Internet of Things Suite integrates and organizes the flow of this data, manages it, analyses it and presents it in a format that helps humans to take appropriate decisions. This highly analyzed and managed data also helps in automation of various processes and operations.

Ayla Networks [8] – Ayla Networks is a cloud-based application enablement platform that offers a cost-effective solution for OEMs to connect any device to the Internet. With an adaptive element for building innovative applications that bridge communications between device, cloud and application, Ayla Networks provides software agents embedded in both connected devices and mobile device applications for end-to-end support. Ayla Networks platform can integrate secure connectivity and data intelligence into products without significant design or business model modifications.

Xively [9] – Xively is a LogMeIn product built on the Gravity Cloud that integrate physical devices and associated data with existing CRM, ERP or other business systems. Xively streamlines development with a PaaS model with searchable libraries of objects and permissions for dozens of languages and platforms, and Xively's RESTful API supports multiple data formats, including JSON, XML and CSV. The open source libraries provide a range of options and integrations with automation platforms such as Salesforce.com and SAP.

ARM [10] – ARM creates sensors, controllers, microprocessors and other types of embedded intelligence for the IoT, enabling objects to sense variables in the environment communicate with other devices and objects and interact with cloud-based applications and other networks. ARM licenses technology to various partners, enabling organizations to add value and differentiate themselves from competitors.

Intel [11] – The Intel® IoT Platform provides an end-to-end platform for connecting the unconnected, allowing data from billions of devices, sensors, and databases to be securely gathered, exchanged, stored, and analyzed across multiple industries. The key benefits are security, interoperability, scalability and manageability by using advanced data management and analytics from sensor to datacenter.

Jasper [12] – Jasper provides features to launch, manage and monetize connected devices and IoT applications. The configurable Jasper Control Board Platform is customizable to suit specific operational needs, business models and requirements across industries and across different geographical locations. Jasper serves IoT needs such as connected cars and enterprise mobility, offering network visibility across devices and real-time monitoring for precise control and deeper insights to drive decision-making.

AllJoyn [13] – AllJoyn enables compatible devices within proximity to recognize one another, communicate and share data across brands, networks and operating systems. AllJoyn is an open-source project of the AllSeen Alliance, providing a common central language to support the Internet of Things and empowering developers and manufacturers with the tools and technologies they need for forward-thinking IoT innovation.

Bosch Software Innovations Suite [14] – The Bosch Software Innovations Suite is modular for advanced flexibility, enabling device management, business process management, and business rules management for the IoT. It integrates seamlessly with existing IT infrastructures for streamlined connectivity and enhanced data analytics. The Bosch Software Innovations Suite is powering the IoT by connecting the four key elements of the ecosystem: People (Users), Things, Enterprises and Partners.

IBM Bluemix^{TM} and Bluemix IoT Foundations^{TM} [15] – IBM Bluemix^{TM} is a PaaS platform for public, private, and hybrid Cloud deployments. Bluemix IoT Foundations is a component of IBM Bluemix that allows to easily integrate IoT devices within the offered PaaS services and to rapidly build new IoT applications. IoT Foundations enables device-to-Cloud registration and security, and scalable communication through transparent MQTT-based services. IoT data flows can be easily combined with other Bluemix services through rich APIs or interactive flow based GUIs.

OpenRemote [32] – OpenRemote is an open-source middleware solution for the Internet of Things, allowing you to integrate any device, regardless of brand or protocol, and design any user interface for iOS, Android or web browsers. Using OpenRemotes cloud-based design tools for developing completely customized solutions, upgrades are streamlined, meaning your devices are literally future-proof.

Arrayent [33] – Arrayent is an IoT platform for connected objects, enabling major brands like Whirlpool, Maytag and First Alert to bring smart, connected devices to consumers. The platform addresses both ends of the product spectrum with both enterprise and consumer apps, coupled with data analytics and a mobile framework for a complete plug-and-play installation at a reasonable cost. The platform scales to support millions of devices.

Echelon [34] – Echelon is an Industrial Internet of Things (IIoT) platform with a full suite of chips, stacks, modules, interfaces and management software for developing devices, peer-to-peer communities and applications delivered via the IzoT Device Stack, IzoT Server Stack and FT 6000 EVK. Echelon is distinct from a consumer IoT platform by addressing the core requirements for the IIoT, including autonomous control, industrial-strength reliability, support for legacy evolution and exceptional security.

Wind River [35] – Wind River has been providing connected intelligence of the IoT-caliber for decades. Wind River provides a foundation for the reliable and efficient operation of IoT networks and connected devices for highly-regulated industries and mission-critical applications. Wind Rivers connects legacy devices to the IoT, Manages sensor data, provide real-time monitoring and analysis, powers sophisticated applications for automotive, aerospace and more, and converge siloed systems.

Contiki [36] – Contiki is an open-source operating system for the Internet of Things, connecting low-cost, low-power microcontrollers to the Internet and enabling rapid, streamlined development. Instant Contiki provides an entire development environment in a single download, and applications are written in standard C. With the Cooja simulator Contiki networks can be emulated before burned into hardware; Contiki runs on a range of low-power wireless devices — most of which can be purchased easily via the Internet. There are a variety of hardware platforms available for free within the Contiki code.

SensorCloud [37] – SensorCloud is a solution from LORD MicroStrain, a company that produces smart, embedded transducers, sensors and sensor networks. SensorCloud provides integrated Big Data analytics, automated alerts and actionable reports for predictive maintenance and streamlined monitoring of connected devices. It includes a unique data storage, virtualization and remote management platform, and SensorCloud supports any device, sensor or sensor network through an OpenData API.

There are a lot of developments of IoT platforms which makes it impossible to cover them all in this section, but some are referred in [38–70].

9.5 IoT Alliances

The Internet of Things applications will be built using complex structures of hardware, sensors/actuators, applications and intelligent devices that need to be able to communicate locally, globally and across industrial verticals.

Alliance for Internet of Things Innovation (AIOTI) [18] – brings together different industries, different sectors (energy, utilities, automotive, mobility, lighting, buildings, manufacturing, well-being, supply chains, cities etc.) and some of Europe's largest tech and digital companies. The AIOTI is an important tool for supporting the policy and dialogue within the IoT ecosystem and with the European Commission and expands activities towards innovation within and across industries. The alliance concentrates on the definition and design of IoT Large Scale Pilots to be funded under the Horizon 2020 Research and Innovation Programme, and will help to build the links and forge the cross-sectorial synergies required for this, as cooperation is crucial for the development of the IoT. The initiative cuts across several technological areas such as smart systems integration, cyber-physical systems, smart networks, and Big Data; and targets SME and IoT innovators to create an open IoT environment. The AIOTI brings together different industries: nanoelectronics/semiconductor companies, Telecom companies, network operators, platform providers (IoT/Cloud), security, service providers and different sectors: energy, utilities, automotive, mobility, lighting, buildings, manufacturing, healthcare, supply chains, cities, etc.

Industrial Internet Consortium (IIC) [19] – brings together the organizations and technologies necessary to accelerate growth of the Industrial Internet by identifying, assembling and promoting best practices. Membership includes small and large technology innovators, vertical market leaders, researchers, universities and governments. This goal of the IIC is to drive innovation through the creation of new industry use cases and test beds for real-world applications; define and develop the reference architecture and frameworks necessary for interoperability; influence the global development standards process for internet and industrial systems; facilitate open forums to share and exchange real-world ideas, practices, lessons, and insights and build confidence around new and innovative approaches to security.

AllSeen Alliance [13] – is a nonprofit consortium dedicated to enabling and driving the widespread adoption of products, systems and services that support the Internet of Everything with an open, universal development framework supported by a vibrant ecosystem and thriving technical community.

The AllSeen Alliance framework is initially based on the AllJoyn^{TM} open source project, and will be expanded with contributions from member companies and the open source community.

Open Interconnect Consortium [20] – is focused on delivering a specification, an open source implementation, and a certification program for wirelessly connecting devices. Membership is currently over 50 members.

Thread Group [21] – has a goal to create the very best way to connect and control products in the home. Thread it's a mesh network designed to securely and reliably connect hundreds of

IPSO Alliance [22] – The Internet Protocol for Smart Objects Alliance is a global forum including many Fortune 500 high tech companies and noted industry leaders. IPSO serves as a thought leader across the board for communities seeking to establish the Internet Protocol (IP) as the network for the connection of Smart Objects.

Eclipse Foundation [23] – Eclipse is a community for individuals and organizations who wish to collaborate on commercially-friendly open source software. Its projects are focused on building an open development platform comprised of extensible frameworks, tools and runtimes for building, deploying and managing software across the lifecycle.

OASIS [24] – is a non-profit consortium that drives the development, convergence and adoption of open standards for the global information society. OASIS promotes industry consensus and produces worldwide standards for security, Internet of Things, cloud computing, energy, content technologies, emergency management, and other areas.

OneM2M [25] – The purpose and goal of oneM2M is to develop technical specifications which address the need for a common M2M Service Layer that can be readily embedded within various hardware and software, and relied upon to connect the myriad of devices in the field with M2M application servers worldwide.

Internet of Things Consortium [26] – Driving adoption of IoT products and services through consumer research and market education.

UPnP Forum [27] – Industry initiative of 1000+ members working to enable device-to-device interoperability and facilitate easier and better home networking.

Bluetooth SIG [30] – Bluetooth Special Interest Group is the body that oversees the development of Bluetooth standards and the licensing of the Bluetooth technologies and trademarks to manufacturers.

Industrial IP Advantage [29] – the goal is to make the most of networking technologies – existing and emerging – that allow integration and the flow of information effortless. This community follows the latest trends, developments, implementation advice and opinions on the use of IP in industrial applications.

ZigBee Alliance [28] – is an open, non-profit association of approximately 400 members driving development of innovative, reliable and easy-to-use ZigBee standards. The Alliance promotes worldwide adoption of ZigBee as the leading wirelessly networked, sensing and control standard for use in consumer, commercial and industrial areas.

9.6 Business Models

Success in the Internet of Things will come for the companies searching beyond the disruptive technology. The Internet of Things technology developments require focusing on complete solutions and integrate the advances in nanoelectronics, cyber-physical systems, and communications with software services, apps, and APIs combined with business models disruption.

For IoT technologies the value will be realized by the alignment of embedded systems technologies, intelligent device communications, network services and IoT infrastructure and application services. The established knowledgeable players in the field and all the newcomers will have to commune and align themselves in ways that will change all the player's business models.

As IoT proliferates, the value creation and value capture is changing and the implications for business model innovation are very important and the well-known frameworks and streamlining established business models are not enough.

Value creation involves performing activities that increase the value of a company's offering and encourages customer willingness to pay. The traditional product companies' business model is creating value by identifying enduring customer needs and manufacturing well-engineered solutions. Competition is based on performance/specifications vs. performance/specifications proposition.

For IoT applications where the devices are connected products are no longer passive participants in the value creation. Over the air updates can create new features and functionality can be provided to the customer on demand or automatically. The ability to track products in use offers the possibility to respond to customer behavior or based on the context. The products are being connected with other products, leading to new analytics and new services for more effective forecasting, process optimization, and customer service experiences.

Though the business models are intermingling today, all major players have operated within established business models that reflected the distinctive competencies at the core of each group. The advent of IoT applications is pushing the boarders between these legacy business models and all the existing emergent players and start-ups as well as the larger IT, software/hardware and network players will have to re-think their strategies.

9.7 Standardization

The IERC previous SRIA addresses the topic of standardization and is focused on the actual needs of producing specific standards. This chapter examines further standardization considerations. Standards are needed for interoperability both within and between domains. Within a domain, standards can provide cost efficient realizations of solutions, and a domain here can mean even a specific organization or enterprise realizing an IoT. Between domains, the interoperability ensures cooperation between the engaged domains, and is more oriented towards a proper "Internet of Things". There is a need to consider the life-cycle process in which standardization is one activity. Significant attention is given to the "pre-selection" of standards through collaborative research, but focus should also be given to regulation, legislation, interoperability and certification as other activities in the same life-cycle. For IoT, this is of particular importance.

Standardization (data standards, wireless protocols, technologies) is still a challenge to more-rapid adoption of the IoT. A wide number of consortiums, standards bodies, associations and government/region policies around the globe are tackling the standards issues.

IoT applications are using wireless protocols and technologies such as wireless MBus, Lora, Weigtless, Sigfox, ZigBee, 6LowPan, DLMS Cosem, COAP, DECT, EnOcean, Bluetooth Low Energy, Hanadu, ULE, X10, Wibree, IPV6, etc. The terms are part of the connected objects' landscape. These technologies are used to cover long and short distances and can be used as wireless radio technologies for M2M and IoT. Some of these technologies are low-data rate, long or short range, undergoing standardization for some or already standards for the others, even being adopted by the market for the most promising ones. Some are suitable for a precise vertical business (Wireless M-Bus for smart meters, KNX for building, ZigBee or EnOcean for the Smart Home), others are more universal (LoRa, Sigfox, Weightless, DECT, ULE, etc.).

A complexity with IoT comes from the fact that IoT intends to support a number of different applications covering a wide array of disciplines that are not part of the ICT domain. Requirements in these different disciplines can often come from legislation or regulatory activities. As a result, such policy making can have a direct requirement for supporting IoT standards to be developed. It would therefore be beneficial to develop a wider approach to standardization and include anticipation of emerging or on-going policy making in target application areas, and thus be prepared for its potential impact on IoT-related standardization.

The sharing of data between a large numbers of devices presents a huge challenge in using and analyse the data. The lack of universally accepted standards for the interchange can affect significantly the deployment of effective IoT solutions. IoT application and solution designers that create IoT platforms may have to make hard choices in choosing between unproven interoperability standards and the legacy ones. The future IoT platforms need to provide federation mechanisms and solutions to assimilate data from multiple vendors and support open API interfaces across platforms.

Today many IoT devices cannot be managed as part of the enterprise network/infrastructure. The enterprise-class network management needs to extend into the IoT-connected endpoints to detect basic availability/uptime of the devices as well as manage software and security updates.

There are many interests from legacy companies to protect their proprietary systems advantages and from open systems supporters trying to set new standards. Multiple standards could evolve that are based on different requirements determined by device class, power requirements, capabilities and uses. This could be an opportunity for platform providers and open source advocates to contribute and influence future standards and can be a threat for interoperability, rapid deployment as many communications and IoT standards already exit.

The development of IoT and IIoT technologies and applications requires increased efforts of governments, organisations and academia to catalyse, coordinate and manage programs that will lead to the use or and development of the effective and general common standards for manufactures.

9.8 Large Scale Deployments and Test Beds

Internet of Things technology and applications are likely to be major drivers of investment and innovation in the communications sector, over the next years, delivering significant benefits to citizens, consumers and industrial end users. These will lead to the introduction of many new and innovative services, e.g. it will be possible to transmit data between many different types of devices, to improve the safety of transportation, reduce the consumption of energy and improve our health. These technological advancements and convergence within the IoT related technologies shape dynamically the emergence of new business models and IoT ecosystems[1], integrating the future generations of applications, devices, embedded systems and network technologies and other evolving ICT advances based on open platforms and standardised identifiers, protocols and architectures.

In this context the next important step to demonstrate and validate the technology in real environments is the deployment of Large Scale IoT Pilots (LSPs).

By bringing together the technology supply and the application demand sides in real-life settings large Scale Pilots will allow promoting the emergent market of IoT and overcoming the fragmentation of vertically oriented closed systems, architectures and application areas that address challenges in different application areas.

While human social and economic activities continue to gravitate towards urban centres, Smart Cities deploy digital and telecommunication technologies to increase administration efficiency and improve the quality of life of their inhabitants.

Cross-domain city challenges in public safety, mobility, lighting and energy efficiency can be addressed by city-centric ecosystems of inter-operable vertical sub-systems. The integration and compatibility of sensors and actuators of connected sub-systems that are often complementary in the public space, in turn, stimulates the development of novel data-driven value-added city services. Due to their high density and ubiquitous nature, connected lighting systems in particular offer the prospect of evolving into platforms acquiring street-level contextual information and delivering city management functions to diverse city stakeholders.

Smart City LSPs need to address challenges in the fields of standard-ization, cyber-security, open data governance and privacy, and validate the novel business models underlying the services provisioned by future city infrastructures.

These IoT Large Scale Pilots have to address technology challenges across the industrial sector verticals and go beyond the M2M, IoT vertical applications developed in recent years, in order to break the silos and to evaluate the real impact of IoT technology across industrial domains.

---

[1]Comprising all stakeholders representing the IoT application value-chain: components, chips, sensors, actuators, embedded processing and communication, system integration, middleware, architecture design, software, security, service provision, usage, test, etc.

The definition of themes needs to have a broader perspective and go beyond the narrower use cases proposed until now since in the future that cross-vertical collaboration and integration will be among the primary benefits of IoT.

In order to evaluate and test the IoT technology Large Scale Pilots themes such as the ones described below should be considered for implementation:

1. Sustainable multi-modal mobility, smart transport infrastructures and energy integration. (Focus on multi-modal mobility including emerging electric mobility, vehicle sharing, autonomous driving/interconnected vehicle/autonomous parking, road infrastructure, and energy integration).

2. Smart Living Environment for Digital Healthcare, general active well-being and healthy ageing. (Focus on IoT enriched citizen-centric indoor and outdoor living environments through wearables and seamlessly integrated IoT in smart buildings and homes, supporting Focus on citizen-centric IoT for creating age-friendly environments, buildings, mobility, cities to foster independent living and active ageing that enables seniors to lead an active life and healthy life not only in cities, but also in rural and remote places – social benefit could be to reduce the pressure to move into the cities, as life in rural areas could be similarly complete and satisfying).

3. Smart Seamless Integrated City Infrastructure. (Focus on citizen-centric IoT services seamless integrated into city infrastructure implemented by city-centric ecosystems of inter-operable vertical sub-systems addressing cross-domain city challenges in public safety, mobility, lighting and energy efficiency).

4. Smart Farming, Agriculture, Food Safety and Security. (Focus on the life cycle of sustainable production of food and the interaction with environment monitoring and health maintenance).

5. Industrial Internet of Things. (Focus on smart production-centric IoT and the automation on the production floor, logistics, autonomous production lines, industrial robots and energy efficiency including the technology to be embedded in manufacturing facilities, production lines, and industrial environments).

6. Environmental Monitoring, Water and Waste Management. (Focus on quality of living and sustainability. IoT services and water, wastewater, environmental parameters monitoring, climate change management and monitoring and integrated services for water resource availability, monitoring of water level variations in rivers, dams the quality of air,

water and monitoring of various chemical substances that can affect the health of humans, animals and plants, but also including the monitoring of other human activity induced pollution such as noise pollution. Incorporation/inclusion of satellite based Earth observation data – as an extreme form of IoT – would be welcome, in particular in combination with data obtained through ground based observation).

7. IoT for critical infrastructure protection, natural disaster prediction and mitigation.

9.9 IoT Innovation Challenges

The report entitled "The Internet of Things Business Index: A quiet revolution gathers pace," [16] found that 30% of business leaders feel that the IoT will unlock new revenue opportunities, while 29% believe it will inspire new working practices, and 23% believe it will eventually change the model of how they operate. The study found that European businesses are ahead of their global counterparts in the research and planning phases of implementing IoT [17]. Manufacturing is the leading sector when it comes to research and implementation of IoT technologies, driven in part by the need for real-time information to optimize productivity [17].

According to the report, the top five concerns that companies have around the IoT are: a lack of employee skills/knowledge; a lack of senior management knowledge and commitment; products or services that do not have an obvious IoT element to them; immaturity of industry standards around IoT; and high costs of required investment in IoT infrastructure. The report suggests that data silos need to be removed and common standards need to be established in order to allow the IoT to scale to a size that will allow it to operate across all markets successfully.

One of the biggest obstacles of using the IoT is the perception that certain products or services do not have any obvious IoT application. The full potential of the IoT will be unlocked when small networks of connected things, from vehicles to employee IDs, become one big network of connected things extending across industries and organizations [16]. Many of the business models to emerge from the IoT will involve the exchange of data; an important element of this will be the flow of information across the networks.

In this context one of the main challenges for innovation in IoT is the complexity of the IoT ecosystem and the vast knowledge required from various fields in order to bring IoT solutions to the market.

Obstacles to innovation (see Figure 9.7) in the IoT field are related to:

- Power management in connected devices, running IoT applications, aimed at reducing their overall energy consumption. Currently, the majority of connected devices run on batteries which have limited shelf life.
- Heterogeneity of technologies, communication protocols and standards used in IoT (wired and wireless).
- Lack of a shared infrastructure, few open platforms based on open-source software, while in contrast, plenty of proprietary technologies/solutions integrated in a silo-based manner.
- Scalable data management, needed to deal with the huge amount of data generated by the increasing number of connected sensors, actuators and "things" in general. Data needs to be gathered and processed in a proper way in order to be able to extract the "smart data" and actually obtain a business value from it.
- Privacy and security in the IoT, guaranteeing secure communications and controlling and deciding who has access to the data is crucial for IoT innovation.
- Data sharing, new alliances and relationships within the IoT ecosystem are necessary to develop a vision in which sharing available data, while preserving its privacy, will drive innovation in IoT.

These obstacles for innovation in the IoT field, as can be seen in Figure 9.7, call for a new concept of federation of IoT platforms. A system of systems that allows enabling the connectivity of a large number of various IoT heterogeneous devices, and is able to collect and aggregate the huge quantities of data generated -while guaranteeing the privacy of the information exchanged-, to support a variety of IoT applications.

Overcoming innovation obstacles to reach the expected impact and potential market for IoT, will require IoT stakeholders to work together within the IoT ecosystem to develop solutions that have greater potential to drive significant business value for both public and private sectors.

Collaboration across the complex IoT ecosystem will bring together a wide range of expertise and capabilities required to create an innovative value chain, including the deployment and implementation of new IoT technology and applications. Much of the potential value will come from moving beyond the proprietary technology silos that largely exist today, and new revenue may come from product and service innovations that could enable growth beyond current products and market segments.

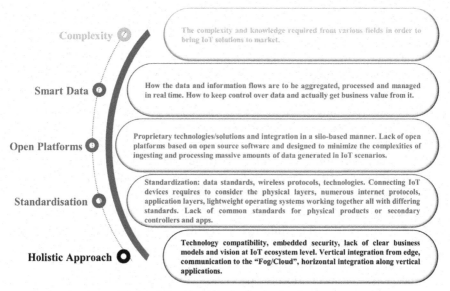

Figure 9.7 Obstacles for innovation in IoT.

Generating new alliances and relationships within the IoT ecosystem is a must to drive innovation together with the availability of an IoT platform, that will allow organizing value creation and help reduce the investment and effort needed to develop new IoT applications and offer novel IoT services.

9.10 Further Developments

The IoT is a disruptive technology in the early stages of market development characterized by innovation, fragmentation, competitive and involving existing and emerging standards. The established technology companies react and adjust to the market developments.

The IoT will leverage, integrate, extend and enhance cloud, smart data, and mobile devices to provide intelligent sensing/actuating devices at the edge. This will generate new applications and use cases that push for new business models and revenue opportunities, while threatening existing industries, markets and products impacting adjacent disrupting trends.

The IoT evolution will generate a shift of the value from devices and components into complete and complex solutions and services embedding opportunities for new value chains, value creation, new business models and new revenue streams and monetizing mechanisms for market participants.

Bibliography

[1] Axeda, http://www.ptc.com/axeda, Accessed 29 May 2015.

[2] ThingWorx, http://www.thingworx.com/, Accessed 29 May 2015.

[3] SAP Internet of Things Solutions, http://go.sap.com/solution/internet-of-things.html, Accessed 29 May 2015.

[4] IoT to Drive Double-Digit IC Sales, 2014, http://electronicspurchasing strategies.com/2014/12/15/iot-drive-double-digit-ic-sales/, Accessed 15 May 2015.

[5] IC Market Drivers—A Study of Emerging and Major End-Use Applications Fueling Demand for Integrated Circuits, 2014, http://www.icinsights.com/services/ic-market-drivers/

[6] McKinsey & Company and the Global Semiconductor Alliance (GSA), May 2015. The Internet of Things: Opportunities and challenges for semiconductor companies. http://www.gsaglobal.org/wp-content/uploads/2015/05/1.-GSA-McK_Report-IoT_Text_Executive-Summary.pdf, Accessed 29 May 2015.

[7] Microsoft Azure, http://azure.microsoft.com/en-us/, Accessed 29 May 2015.

[8] Ayla Networks, http://www.aylanetworks.com/platform/aylas-iot-cloud-fabric, Accessed 29 May 2015.

[9] Xively, https://xively.com/, Accessed 29 May 2015.

[10] ARM, http://www.arm.com/markets/internet-of-things-iot.php, Accessed 29 May 2015.

[11] Intel, http://www.intel.com/content/www/us/en/internet-of-things/iot-platform.html, Accessed 29 May 2015.

[12] Jasper, http://www.jasper.com/, Accessed 29 May 2015.

[13] AllJoyn, AllSeen Alliance, https://allseenalliance.org/, Accessed 29 May 2015.

[14] Bosch IoT Suite, https://www.bosch-si.com/products/bosch-iot-suite/benefits.html, Accessed 29 May 2015.

[15] IBM Bluemix, https://console.ng.bluemix.net/, Accessed 29 May 2015.

[16] http://www.arm.com/files/pdf/EIU_Internet_Business_Index_WEB.PDF, Accessed 15 May 2015.

[17] http://news.techworld.com/networking/3476043/arm-report-businesses-look-make-money-through-internet-of-things-revolution/ Accessed 15 May 2015.

[18] The Alliance for Internet of Things Innovation (AIOTI), http://www.aioti.eu/, Accessed 29 May 2015.

[19] Industrial Internet Consortium (IIT), http://www.iiconsortium.org/about-us.htm, Accessed 29 May 2015.

[20] Open Interconnect Consortium, http://openinterconnect.org/, Accessed 29 May 2015.

[21] Thread Group, http://threadgroup.org/, Accessed 29 May 2015.

[22] IPSO Alliance, http://www.ipso-alliance.org/, Accessed 29 May 2015.

[23] Eclipse Foundation, http://www.eclipse.org/, Accessed 29 May 2015.

[24] OASIS, https://www.oasis-open.org/, Accessed 29 May 2015.

[25] OneM2M, http://www.onem2m.org/, Accessed 29 May 2015.

[26] Internet of Things Consortium, http://iofthings.org/#home, Accessed 29 May 2015.

[27] UPnP Forum, http://www.upnp.org/, Accessed 29 May 2015.

[28] ZigBee Alliance, http://www.zigbee.org/, Accessed 29 May 2015.

[29] Industrial IP Advantage, http://www.industrial-ip.org/, Accessed 29 May 2015.

[30] Bluetooth SIG, http://www.bluetooth.com/Pages/about-bluetooth-sig.aspx, Accessed 29 May 2015.

[31] O. Vermesan, P. Friess, P. Guillemin, H. Sundmaeker, et al., Internet of Things Strategic Research and Innovation Agenda, Chapter 3 in *Internet of Things – From Research and innovation to Market Deployment*, River Publishers, 2014, ISBN 978-87-93102-94-1.

[32] OpenRemote, accessed online at http://www.openremote.com/

[33] Arrayent, accessed online at http://www.arrayent.com/

[34] Echelon, accessed online at http://www.echelon.com/izot-platform

[35] Wind River, accessed online at http://iot.windriver.com/#!/iot/

[36] Contiki, accessed online at http://www.contiki-os.org/

[37] SensorCloud, accessed online at http://www.sensorcloud.com/

[38] mnubo, accessed online at http://mnubo.com/

[39] Oracle Internet of Things, accessed online at http://www.oracle.com/us/solutions/internetofthings/overview/index.html?ssSourceSiteId=ocomcn

[40] Swarm, accessed online at http://buglabs.net/products/swarm

[41] Etherios, accessed online at http://www.etherios.com/

[42] ioBridge, accessed online at https://www.iobridge.com/

[43] Zatar, accessed online at http://www.zatar.com/

[44] Sine-Wave, accessed online at http://www.sine-wave.com/platform

[45] EVRYTHNG, accessed online at https://evrythng.com/

[46] Exosite, accessed online at http://exosite.com/

[47] Marvell, accessed online at http://www.marvell.com/solutions/internet-of-things/

[48] Carriots, accessed online at https://www.carriots.com/

[49] Arkessa, accessed online at http://www.arkessa.com/

[50] GroveStreams, accessed online at https://grovestreams.com/

[51] CeNSE by HP, accessed online at http://www8.hp.com/us/en/hp-infor mation/environment/cense.html#.VWw2EkYw-uF

[52] Nimbits, accessed online at http://www.nimbits.com/index.jsp

[53] Open.Sen.se, accessed online at http://open.sen.se/

[54] Paraimpu, accessed online at https://www.paraimpu.com/

[55] NewAer, accessed online at http://newaer.com/

[56] ThingSpeak, accessed online at https://thingspeak.com/

[57] Yaler, accessed online at https://yaler.net/

[58] XobXob, accessed online at http://www.xobxob.com/

[59] Linkafy, accessed online at http://www.linkafy.com/

[60] Revolv, accessed online at http://revolv.com/

[61] Wovyn, accessed online at http://www.wovyn.com/

[62] Microsoft Research Lab of Things, accessed online at http://www.lab-of-things.com/

[63] InfoBright, accessed online at https://www.infobright.com/index.php/int ernet-of-things/

[64] 2lemetry, accessed online at http://2lemetry.com/

[65] InterDigital, accessed online at http://www.interdigital.com/iot#.U4fvQ_ ldXy0

[66] Superflux Internet of Things Academy (IoTA), accessed online at http://www.superflux.in/work/iota-phase2

[67] HarvestGeek, accessed online at http://www.harvestgeek.com/

[68] MediaTek Labs, accessed online at http://www.harvestgeek.com/

[69] StreamliteLTE, accessed online at http://www.sequans.com/products-solutions/streamlitelte/

[70] RIoTboard, accessed online at http://www.riotboard.org/

Index

Lightning Source UK Ltd.
Milton Keynes UK
UKOW06n1213100117
291770UK00005B/68/P